AF446774

Rural Households in Emerging Societies

The constantly altering circumstances of rural life in sub-Saharan Africa have brought with them both successes and failures. The essays in this volume examine the various pressures and inducements to changing resource-use patterns faced by rural households, and explore the two-way causal relationship between technology and technological change on the one hand and other key elements of rural change – demographic, environmental, economic, social, and political – on the other. Contemporary approaches to the introduction of technical innovations are examined, and new approaches are proposed. Through case studies of particular communities, the wide-ranging impacts of past experiences are assessed, and the causes and consequences of indigenous initiatives are explored.

Margaret Haswell is an Independent Consultant in Rural Development.

Diana Hunt is Lecturer in Economics at the School of African and Asian Studies, Sussex University.

Rural Households in Emerging Societies

Technology and Change in sub-Saharan Africa

Edited by
Margaret Haswell
and
Diana Hunt

BERG
Oxford/New York
**Distributed exclusively in the US and Canada by
St Martin's Press, New York**

First published in 1991 by
Berg Publishers Limited
Editorial offices:
165 Taber Avenue, Providence, RI 02906, USA
150 Cowley Road, Oxford OX4 1JJ, UK

British Library Cataloguing in Publication Data

Rural households in emerging societies: technology and
    change in sub-Saharan Africa.
    1. Africa south of the Sahara. Rural communities. Social conditions
    I. Title   II. Haswell, Margaret   III. Hunt, Diana
    307.720967

Library of Congress Cataloging-in-Publication Data

Rural households in emerging societies : technology and change in
    sub-Saharan Africa / edited by Margaret Haswell & Diana Hunt.
        p.   cm.
    Includes index.
    ISBN   0–85496–730–3
    1. Rural development–Africa, Sub-Saharan   2. Rural development–Africa,
Sub-Saharan–Case studies.   3. Family farms–Africa, Sub-Saharan.
4. Family farms–Africa, Sub-Saharan–Case studies.
I. Haswell, Margaret Rosary.   II. Hunt, Diana, 1942–
HN780.Z9C6722   1991
307.1′412′0967–dc20   90–26657
                                                              CIP

.

Printed in Great Britain by Billing & Sons Ltd, Worcester

Contents

Contents

vi

List of Tables, Figures and Maps

Figures

Notes on Contributors

Domien **Bruinsma** trained as a food technologist at the Agricultural University of Wageningen, The Netherlands. He spent six-years in Mozambique as a governmental technology consultant. Presently, he is attached to the section *Food and Nutrition* of the International Agricultural Centre, Wageningen, where he is in charge of advisory tasks and the planning and design of a post-graduate course on food quality assurance.

Robert **Chambers** went on from his original work in public administration to include other aspects of rural development including national policy, settlement schemes, rural refugees in Africa and agrarian change in South Asia. His interests include seasonality, rapid rural appraisal, canal-irrigation management, agricultural research, social and agro-foresty, the role of NGOs in relief and rural development, and professionalism and the nature of rural poverty. He is a fellow of The Institute of Development Studies at the University of Sussex.

Angela **Cheater** is professor of sociology at the University of Zimbabwe. She has researched extensively in Zimbabwe among black freehold farmers, as well as among fishermen and industrial workers. She has also researched in China on comparative studies.

Josephine Wanja **Harmsworth** trained as an anthropologist and went to live in Uganda in 1960. She has had a varied work experience, chiefly as a socio-economist, and is now an independent consultant.

Margaret **Haswell** is an independent consultant specialising in village survey techniques, social infrastructure, monitoring and evaluation of operational problems. A former Oxford University lecturer in agricultural economics and former Fellow of St Hugh's College, she has worked in East and West Africa, South and Southeast Asia, the Caribbean and the South Pacific, while continuing to research on agrarian change in a middle river area of The Gambia.

Diana **Hunt** is an economist based at the School of African and Asian Studies, University of Sussex. She has long-standing research and consultancy experience in the fields of agricultural and rural development in Africa. She has conducted lengthy periods of field research in Uganda (on the organisation and impact of small farm credit programmes) and in Kenya (on the impact of government development policy on rural households of varying socio-economic status in semi-arid areas).

Robert **Nout** trained as a food scientist at the Agricultural University of Wageningen. He spent twelve years in Mali, Nigeria and Kenya, involved in university manpower training of food technologists. Presently, he is a senior lecturer in Food Microbiology at the Agricultural University of Wageningen.

Paul **Richards** is reader in Anthropology at University College, London. He is currently working on social aspects of rain-forest conservation in Sierra Leone, and has recently completed a comparative study of the impact of rice research in South Asia and West Africa with Michael Lipton and Adam Pain.

Paul **Starkey** is an independent consultant, specialising in animal traction. He has been a Technical Co-operation Officer in Sierra Leone, where he initiated a national animal traction programme. He is currently Technical Adviser of the West Africa Animal Traction Network, and an Honorary Reseach Fellow at the Centre for Agricultural Strategy in the University of Reading.

Camilla **Toulmin** is currently Director of the Drylands Programme at the International Institute for Environment and Development, London. The main concerns of this programme are research, networking and support to non-governmental organisations in Africa through information and training activities. An economist by training, she carried out her thesis field work in Mali.

Jan Douwe **van der Ploeg** trained as an agronomist, and worked for fifteen years in Latin America and Africa, including Guinea Bissau. He then took a PhD in Social Sciences and now teaches at Wageningen University, The Netherlands, while continuing to research on agrarian change in southern Europe, Latin America and Africa.

Editors' Preface

All the papers presented in this book, with one exception, that by Paul Richards, were first discussed at a workshop held at the School of African and Asian Studies, University of Sussex, in September 1989. Underlying the broad title of the workshop was the thesis that there is a two-way relationship between technology and technological change on the one hand, and other elements of rural change – demographic, economic, social, cultural and political – on the other. The focus throughout our discussions was upon *change*, the experiences of rural households as social and economic actors in the constantly altering circumstances of rural life.

The book argues for greater caution in predicting the advent of any apparently unambiguous phenomenon called 'rural development'. It also suggests various changes in approach both to analysis and to research and extension, while through case-studies of particular communities it assesses the wide-ranging impacts of past experiences of technical change on rural households.

In addition to those whose papers appear in this volume, we should like to acknowledge the contributions, through submission of papers and/or discussions, of Derycke Belshaw, Tony Binns, Sally Chilver, Alison Evans, Donald Funnell, Jan Kuyper, Robin Mearns, Andrew Millington, Phil O'Keefe and Housainou Taal, who also played a role as one of the instigators of the project. We should like to thank Carol Brown, formerly of the School of African and Asian Studies, for her much valued administative and secretarial support, and Amanda Davey for her artwork for some figures.

We also wish to express our gratitude to St Hugh's College, University of Oxford, The Nuffield Foundation and the School of African and Asian Studies, University of Sussex, which provided financial support both for the workshop and for the preparation of the book.

We are grateful to Professor T. T. Isoun, Editor, The African Academy of Sciences, for permission to reproduce an abbreviated version of Paul Richard's paper, which was published in the first issue of their *Discovery and Innovation*.

Finally, we should like very warmly to thank Sally Chilver for her invaluable editorial assistance and for so much generous help.

Margaret Haswell
Diana Hunt

1

Introductory Chapter

Margaret Haswell and Diana Hunt

We begin with a brief explanation of our title: the phrase 'emerging societies' refers to the emergence of previously relatively isolated societies into a context in which they face both pressures and inducements to enter into a wider and deepening network of market relations.

This book is about the experiences of rural households as farmers, and as social and economic actors more generally, in the face of a constantly changing physical, demographic, political and market environment. Our focus is on rural households in Africa south of the Sahara, a vast region which contains some of the most difficult production environments in the world, as well as other, better favoured areas. Indeed, a key feature of this region is its diversity, something of which we try to indicate in what follows. But a study of this kind cannot hope to be comprehensive, either in terms of issues reviewed or sub-regions and societies analysed. Instead, we have aimed to develop certain broad themes through the medium of contributions that are issue-oriented and through surveys and case-studies.

Underlying the broad sub-title 'Technology and Rural Change in sub-Saharan Africa' is the thesis that there is a two-way relationship between technology and technological change on the one hand, and other key elements of rural change – demographic, economic, social, cultural and political – on the other. Among the most important of the variables involved in this complex process of change are: household size and composition, child school enrolment, the farm enterprise mix, the full household economy range of economic activities, the stock of natural resources, the role of livestock within household and rural economy, the man-made physical infrastructure (most notably, roads, health facilities, water supplies, also electricity supplies, telephone connections, etc.), and, of course, the actual production methods used in different branches of rural economic activity – whether it be in food processing, cooking, obtaining water, farming, non-farm commodity production or any other branch of the household productive sphere. Among important examples of the two-way causal relationship between technology and technical change and these other variables, and among the changes that are directly or indirectly generated through both types of causal relationship, we may note:

1   The potential impact of production technology on household size and the impact of household size and composition on viable production technology.
2   The impact of technical change on intra-household divisions of labour and income, as well as the significance of the latter as determinants of the scope for technical change.
3   The impact of technical change on the physical environment, and the role of changes in the physical environment as determinants of need for technical change.
4   The role of household goals and priorities in determining both the need for and the acceptability of technical change, as well as the impacts of technical change on household values and priorities.
5   The impact of different types of production technology and technical change on attitudes towards placement of children in school, and the impact of formal education on attitudes and reactions to technical change.
6   The relationship between technical change and the need for credit, and the impact of indebtedness on households' ability and willingness to innovate.
7   The impact of technology on the gender division of labour, and the impact of the gender division of labour on willingness and ability to adopt technical innovations.

Causal relationships such as these and their consequences are, however, as yet only imperfectly understood, and these and related issues are explored in the chapters that follow.

The editors invited the contributors to this volume to explore in their papers one or more of three themes which seem to us to have a particular significance for change in production processes in contemporary rural Africa. These themes, which the reader will find taken up with varying emphases in the following chapters, are:

1   The need in sub-Saharan Africa for rural technologies that are both environmentally sustainable and can support rising popu-

lations at higher income levels; as well as the need to recognise the value of local knowledge and initiatives in developing such technologies.

2 The need to plan and appraise technical innovation in the context of full household economic systems.

3 The search for (varying degrees of) autarky by African rural producers in the face of market failure, cash constraints and other pressures.

In addition, however, at the workshop at which these papers were discussed, it became clear that a number of other common themes were also emerging. We review these briefly in the following paragraphs.

The focus throughout the collection is on *change* rather than on that value-laden but nebulous concept 'development'. This is because the constantly altering circumstances of rural life in sub-Saharan Africa have brought with them both successes and failures and not the steady progression from 'backward' to 'developed' state that was once too naively anticipated. This more complex reality is now generally acknowledged. But the assessment of which changes fit which label inevitably varies, both among local people and 'outsiders', depending, *inter alia*, on values and personal preferences.

Throughout this volume emphasis is also placed on the need to develop, as far as possible, a holistic approach to the *analysis* of rural change; to explore such change in the round (including at the household level, in the context of the full socio-economy and not just isolated fragments thereof) in an attempt to comprehend and anticipate the broader and longer-term ramifications of particular innovations, whether these be endogenous or exogenous. Such holism entails the need to analyse the *causes* of the constraints on, and impact of, change at the household, community and regional levels.

It is difficult, too, to overemphasise the importance of endeavouring to analyse processes of change *as they occur over time*, rather than on the basis of a single snap-shot or one-off survey, valuable as the latter may sometimes be. For without such a perspective, an understanding of many of the tensions, and cycles, involved in long-term rural change is inevitably diminished, with the result that it becomes difficult to make effective predictions into the future. (The latter must in any case always be tentative.)

The significance of analysing change over time is brought out particularly strongly in Chapters 6 and 8 by van der Ploeg and

Haswell, respectively. In both these case-studies the authors show how a period of change, characterised by rising welfare and output, can be followed, for a variety of reasons, by one of decline. Haswell also demonstrates clearly the need for holism: the need to recognise that overcoming one particular production constraint – in this case access to the mangrove swamp for rice cultivation – may, if introduced without consideration of the wider implications, generate other, possibly more serious, constraints in the future. In one example a classic instance of the operation of the 'law of the commons' has led to the denuding of the mangrove forest and, consequently, saltwater encroachment into the paddy fields. Thus the full cycle of change instigated by the initial intervention – the construction of causeways into the swamp – has so far been increasing rice output as the cultivated area expands, followed by a decline as yields fall. This decline will in turn almost certainly generate further change: probably the replanting of the mangrove forest, this time for industrial purposes.

Some of the other case-studies also either indicate analogous chain-reactions or prompt the question of whether, while not yet apparent, they can be anticipated in the future. Thus, Harmsworth's study (Chapter 9) of the impact of small farm tobacco production in north-west Uganda notes the impact of flue-curing on local woodfuel supplies. In this case, the instigators of change recognised the need to match rising demand with rising supplies but, for a variety of reasons, have not yet been successful in securing these. Toulmin's account (Chapter 7) of indigenous well-digging in central Mali to secure from Fulani herdsmen cattle dung for millet fields raises the question of what in the future will happen to the water-table and, hence, the viability of this newly developed agricultural system – itself a response to adverse climatic change.

Outside agencies influence rural production environments both directly and indirectly – directly through the provision of infrastructure, knowledge and inputs; indirectly through the provision of health care and formal education facilities for children. Both these services in turn influence both labour supply – the former positively, the latter, in quantitative terms, negatively – and the demands made upon the production environment: more mouths to feed, school fees to be met. Examples of the impact of exogenously introduced change are provided in all of the case-study chapters. Two types of such intervention, which are not directly geared to the promotion of technical change in production *per se*, are particularly worthy of note, partly because they have tended to be underemphasised in the past.

4

One such intervention is the development of transport facilities (see Chapters 8, 9 and 10). The opening up of access to new productive resources and new market opportunities can bring dramatic changes to rural areas. In this respect one is led to speculate, too, on what changes will occur in the future to the community of Kala (see Chapter 7) should improved communications draw its members more fully into wider market networks.

A second external intervention, which can be shown to have important implications – whether actual or potential – for the communities described in the case-studies, is the advent of formal education opportunities for children. The draining away of child labour from its established productive role has important organisational, technical and welfare implications for the household economy. The children's roles may not be energy-intensive, but none the less are often crucial. How, for example, do you produce substitute bird-scarers? In practice, the loss of children from bird-scaring may entail a change in the farm enterprise mix in order to minimise the risk of this type of crop loss. Meanwhile the loss of children from herding also entails change – either organisational and/or in the composition of household assets. All these adaptations have implications for household welfare. Harmsworth (Chapter 9) shows how households adopting new, more modern and profitable, but labour-intensive, forms of production may be reluctant to educate children in order to secure their labour.

A third exogenous change, which is also not directly geared to production, is, of course, the expanded provision of health care with its consequent implications for population growth. Demographic change due both to natural reproduction and migration is widely recognised as having important impacts on rural communities in Africa and on the natural environment on which they still rely for survival. Yet there are still areas where population pressure is not yet a practical issue (as in some of the case-studies described in this volume). In these cases there is a need to try to predict when critical population levels are likely to be reached, and to begin to raise local awareness of the implications and of the need for appropriate adaptation to minimise the stresses that will otherwise be caused.

However, while rising population numbers are now most commonly seen as a problem in discussions of rural change in Africa, what emerges most markedly in the case-studies is a continuing recognition of the importance of having a large labour force within the household. Toulmin, van der Ploeg, Haswell and Harmsworth all demonstrate the importance of household size in generating and maintaining economic and social success. This is seen as the only reliable means of securing access to what, given presently affordable production technologies, remains a key productive resource, and one that is invariably in short supply at periods of peak farm activity. Even the technologically more advanced households described by Cheater (Chapter 10) have large numbers of productive members by western standards.

One piece of now increasingly widely received conventional wisdom, to which the authors all subscribe, is that rural households innovate. Their own production technologies are *not* static, but rather subject to experimentation and change (see Chapters 2, 3 and 6 in particular). In this regard, however, some of the studies also draw attention to two important, related issues: the need to understand the households' own criteria of efficiency, and to appreciate the reasons for, and significance of, their preferences for significant degrees of autarky in their production activities – albeit varying between communities. With respect to efficiency, it is important to recognise, *inter alia*, that a productive resource may simultaneously generate multiple products each with use values, and thus its performance should not be judged solely in terms of one of these (see e.g. Chapter 10 on the multiple products of stoves; and Chapter 4 on the joint products of labour).

The search for autarky can be viewed in various ways. Cheater (Chapter 10) interprets this search as an attempt to balance energy sources in production with a view to achieving affordable and sustainable energy supplies in the face of risks of market failure. Van der Ploeg (Chapter 6) emphasises that well-designed, outside interventions must also recognise the desirability and rationality of autarky, and help rural households to sustain those forms of autarky that are central to their particular production structures.

Energy sources for household production provide a focus for four of the contributions to this volume (Starkey, Haswell, Cheater, Bruinsma and Nout). The view held by all the contributors is that there is a need to encourage the use of energy sources that are adapted to the needs and resource availabilities of rural households, and that, where possible, these sources should fall within the control of the households themselves. In many parts of sub-Saharan Africa animal traction holds considerable promise in this respect (Starkey, Toulmin, Haswell and Cheater). But we should be wary of assuming that comparable intermediate or appropriate technologies are equally readily available to meet all rural needs (see Bruinsma and Nout, Chapter 11, on food processing technology).

This book does not offer concrete solutions to problems posed by rural change. Rather, it argues for greater caution in predicting the advent of any apparently unambiguous phenomenon called 'rural development', and it suggests various changes in approach, both to analysis (Chapter 4) and to research and extension (Chapter 3) which might help outsiders to contribute more effectively to promoting effective and sustainable changes of the kinds desired by rural communities themselves. It is not without significance that the final chapter (Bruinsma and Nout on food processing technology) concludes with a question. We still have much to learn – and, indeed, our knowledge can never be complete.

Part I begins with a reminder from Paul Richards of the innovativeness of small farmers in the third world, and a discussion of ways in which scientists may not only recognise, but collaborate with, and actively promote, indigenous innovation. Chambers and Toulmin take up this theme and propose in greater detail a new approach to technology development for low-income farmers in those complex, diverse and risk-prone farming areas of the tropics that are unsuited for the green revolution technology. Hunt then argues for the importance of adapting this approach to give equal weight to all productive activities undertaken by rural households, not just farming. Lastly Starkey provides an overview of the uses of animal traction in sub-Saharan Africa. Part II presents the various case-studies referred to above.

PART I

Key Issues

2

Experimenting Farmers and Agricultural Research

Paul Richards

A Case of Indigenous Experimentation

Mende farmers in Sierra Leone have a word, *hungoo*, for experiment. Consider a household short of rice for planting. The bushel of seed borrowed from friends looks doubtful, so it is necessary to carry out a test. A hundred or so grains and a fistfull of moisture-retentive soil are thoroughly mixed together and wrapped in a leaf the size of a dinner plate from the *mbondar* tree (*Mitragyna stipulosa*). This little parcel is kept for three or four days and then opened. The number of germinated seedlings are counted. If the failure rate is more than about 30 per cent the batch of seed from which they were taken is rejected. This is *hungoo*.

Take another case. Harvesting of rice in Mende country is done panicle by panicle with a knife. The owner of the farm may ask the harvesters to take special care to separate out any off types. Mostly, these are familiar varieties, accidentally mixed at planting time or carried to the farm by birds or passers-by. Sometimes, however, the harvesters come across a rice plant (perhaps a spontaneous cross) which, according to appearances, they judge to be a new variety. If it looks interesting – a broad category as far as Mende farmers are concerned, since varieties are chosen as much for their looks or curiosity value as for their utility – the seed will be set aside for planting in a small trial plot the following year. The site selected is sometimes a fertile and favoured spot near the farm hut. Alternatively, the chosen plot will be aligned to run across the boundary

This is an abbreviated version of a paper first published in 1989 in *Discovery and Innovation* (Journal of The African Academy of Sciences, PO Box 14798, Nairobi). We are grateful to the editor of the journal, Professor T. T. Isoun, for permission to reproduce this version.

between two distinct soil types. The amount of seed planted is often carefully measured by volume and the harvest measured in the same vessel (perhaps a cup or calabash). Sometimes farmers will plant equal amounts of a familiar and unfamiliar variety along side each other to test how well they do in comparable conditions. This is also *hungoo* (Johnny, 1979; Richards, 1986).

Mende farmers believe that trials and tests of this sort are a long-established element in indigenous culture, even if from time to time the better-travelled among them helpfully point to parallels with trials they have seen on government or university experimental farms (Johnny, 1979). European agricultural officers were reporting experimentation on local farms in the 1930s and 1940s and thought it perfectly feasible that this pre-dated colonial rule (Richards, 1986). Until recently, however, there has been little documentation of on-farm experimentation by small-scale farmers, either in Sierra Leone or elsewhere in the tropical world. Challenged by colleagues at Wageningen to cite references in the literature, Box (1989) was able to point to only one paper (Johnson, 1972) prior to the late 1970s. (Although he might also have referred his colleagues to material on farmer experiments in an earlier, seminal study of agriculture among the Zande of southern Sudan by the Belgian agronomist de Schlippe, first published in 1956.)

Traditional Agriculture, Innovative Cultivators

Older literature readily conceded that agricultural practices were well adapted to environmental conditions in pre-industrial societies. But these practices were often thought of as 'traditional' and static – as if arrived at by happy accident at some early point in the evolutionary sequence, and then copied without further thought generation after generation.

These earlier perceptions are now thoroughly outmoded, for two reasons. First, the facts of agrarian history frequently imply a greater dynamism than the older conventional wisdom was prepared to allow (Boserup, 1965; Rhoades, 1989). How else, as Bunch (1989) notes, was it possible for two New World crops – maize and cassava – to become established as major staples throughout tropical Africa in the 450 years since they were first introduced by the Portuguese? Second, there has been a crucial shift in the way the concept 'traditional' is regarded by historians and social scientists. It is now common to stress that tradition need not be timeless or static (Hobsbawm and Ranger, 1983). Some

12

traditions are quite clearly recent inventions. In other cases even to maintain the appearance of tradition in a changing world requires ceaseless invention and adaptation. Some of these newer perspectives have rubbed off on agriculturalists, and it no longer causes surprise to find that traditional cultivators are tirelessly inventive. (Mende farmers contrast chemical fertilisers with what they describe as their *traditional* fertiliser, *Calpogonium mucunoides*, a leguminous creeper first introduced in the 1930s as part of a series of green manure experiments!) In recent literature the experimenting, innovative, adaptive peasant farmer is now accepted as the norm, not the exception (Atteh, 1980, 1984; Biggs, 1980; Biggs and Clay, 1981; Uzozie, 1981; Millington, 1982; Budelman, 1983; Reij, Turner and Kuhlman, 1986; Richards, 1986; Altieri, 1987; Lightfoot, 1987).

Where reservations are still voiced they generally relate to two points. Surely (unlike the scientist) small-scale farmers only ever experiment for practical reasons? And is it not the case that scope for farmer experimentation is severely limited by the scale of present difficulties in hazard–prone tropical environments? On the first point there is good evidence that so–called 'traditional' cultivators are quite capable of experimenting for intellectual as well as practical reasons, or even for fun – an argument first developed in detail in the work of the anthropologist Lévi–Strauss (1966). On the second point, recent research, from both Africa and India, concerning adaptive responses to drought and famine tends to suggest that the poor farmer's faith in the value of experimentation is strengthened rather than undermined by adversity (Juma, 1987; Vaughan, 1987). De Schlippe (1956), with his usual percipience, was one of the first to provide evidence on this point, recording that experiments by Azande cultivators tended to increase in number and complexity after a poor harvest. Of the examples he documents, some of the most ingenious were undertaken by women.

Farmer Experimentation and Agricultural R & D

A number of agriculturalists have recently begun to explore the idea that indigenous inventiveness provides an effective point of contact between scientific research in tropical agriculture and potential groups of small-farmer clients (Rhoades and Booth, 1982; Chambers and Ghildyal, 1985; Rhoades et al., 1985; Chambers and Jiggins, 1986; Ashby, 1987; Sanghi, 1989). The case advanced is based on both ecological and equity considerations. Conventional agricultural research has so far found great difficulty in reaching poor farmers. Typically, Green Revolution innovation packages are too standardised or too expensive to work well for low-resource farmers in high-risk environments (Gupta, 1989; Richards, 1985). In effect this is to exclude the majority of farmers in rain-fed agricultural districts in India and sub-Saharan Africa. It has recently been estimated, for example, that Green Revolution innovations in rice are only appropriate to the 30 per cent of Asia's rice-growing lands where full water control is possible (Greenland, 1984).

The International Agricultural Research Centers and other research projects in tropical agriculture are beginning to develop a new range of innovations, which look increasingly appropriate to the needs of typical farm households in resource-poor, high-risk environments. Many of these developments derive from principles and expand upon techniques, such as intercropping, already well established in traditional agriculture (Igbozurike, 1971, 1977; Steiner, 1982; Clawson, 1984; Altieri, 1987). Alley farming, for example, ingeniously recombines a number of themes and motifs dear to the heart of the subsistence cultivator, e.g. polyculture, browse for small ruminants, protection against leaching and soil erosion, and use of nitrogen-fixing plants to assist rapid recovery of soil fertility. None of these by themselves will be new to small-scale cultivators in the tropics. But the combination is new and highly complex. The problem with alley farming and similar innovations is, given this complexity, how are they to be tested and fine-tuned for specific on-farm application?

This is where farmer experimental skills may have an important role to play. Sumberg and Okali (1989) identify a minimum of seven biological variables to be analysed in connection with the tree component in alley farming systems. When the other crops and socio-economic factors are also taken into account, this far exceeds in complexity anything manageable by classic research methods. Perhaps, then, there is a case for a totally different approach. If small-scale farmers are regular experimenters, then why not set them up at an early stage with some of the inputs and resources available to the researchers, and a general outline of the theme to be pursued. Taking alley farming as a case in point, the aim would be to encourage farmers to play around with the basic idea in whatever way they saw fit, with the researchers concentrating on trying to make sense of what different farmers were doing and why. Sumberg and Okali (1989) envisage that as a result of this interactive process researchers would be in a much better position to penetrate

14

the initial complexity and home in on key hypotheses more amenable to investigation by established on-station methods.

This proposal is typical of the debates now taking place within the emergent field Robert Chambers has termed 'complementary methods' in agricultural research (Chambers, Pacey and Thrupp, 1989: XIII). The choice of term reflects the aim of supporting and extending rather than replacing conventional experimental methodologies in agriculture. A recent conference at the Institute of Development Studies at the University of Sussex brought together thirty-eight papers (now published as *Farmer First*, 1989) on the subject from a more or less evenly balanced group of social and biological scientists.

Three overarching themes emerged from the papers presented. The first was the notion of complementary methods as a means of developing complex agro-ecosystem innovations *in situ*.

A second theme was that complementary methods should be a way of supporting and perhaps further enhancing indigenous experimental and adaptive capacities.

A final general theme was the notion that farmer experimentation, in addition to being a potentially useful resource to be tapped by research and development agencies, was important and valuable for its own sake. There are good educational and cultural reasons for according experimenting farmers recognition as members of the scientific community. Science is not the narrow preserve of any particular social group or historical tradition. Its values are to be applauded wherever they are to be found.

Examples of Complementary Methods

Farmer Experiments as a Guide to Fruitful Research Agenda

It is a well-understood point of scientific method that understanding the problem is often much more difficult than finding the answer. In its haste, much tropical agricultural development has tried to make do with other people's answers. Technology transfer, however, generally proves ineffective for poor farmers in hazardous tropical environments. This is often because the complexity of the ecological and social issues involved is greater than anything addressed by researchers hitherto. There is little alternative, therefore, to specially designed research for the region in question. But how are high-cost researchers to be cost-effective in complex low-resource environments? One answer is to merge with the background – to research on topics already being addressed by farmers in the region, perhaps according to methods with which farmers are already familiar, with the eventual aim of handing over responsibility for further developments to farmers themselves.

The first issue, then, is to understand farming problems through farmers' eyes. There are various techniques for setting about this task. They include analysing agricultural activities from a participant observation perspective, mapping farm fields and village environments using local categories for soils and vegetation, or organising meetings with farmers to define and refine problem agenda (Edwards, 1987, Chambers and Toulmin, this volume, Chapter 3). But in other cases more unusual methods have been proposed. Box, working with cassava cultivators in the Dominican Republic, has found that the farmer's biography often contains information vital to understanding the pattern of on-farm experimentation (Box, 1989). Rocheleau (1987) describes some of the ways in which she set about stimulating local debate concerning vegetation dynamics. She collected oral historical information and asked informants to envisage what the landscape might look like when they or their children reached old age. Robert Rhoades, an anthropologist with the International Potato Center in Peru, has described the humbling experience, for researchers, of trying to run a small farm village-style with village resources. He also found 'conversation pieces' important for stimulating discussion with farmers. In his case this involved hauling from farm to farm a sack of potatoes containing as many local varieties as he could find (Rhoades, 1989).

Refining relevant and feasible research agenda from material of this sort, in many cases, will depend on putting together a number of methods in a coherent sequence, such as those described by Chambers and Toulmin (this volume, Chapter 3).

Farmer Participation in Varietal Trials

Varietal trials are the stock-in-trade of much agricultural research. The idea of involving farmers in trials is not a new one. And, as recent work has shown, there are a number of ways of strengthening the part that farmers can play in the organisation, execution and modification of such trials.

In Colombia a panel of small-scale farmers, chosen from those considered by the local community to be experts in the crop in question, have been involved in selecting and planting out advanced breeding lines of cassava and beans to provide plant breeders with guidance about where to concentrate their energies in the future (Ashby, Quiros and Rivera, 1989). The exercise was useful because it revealed that farmer assessments were often at variance with the expectations of scientists about what varieties would best appeal. But Ashby et al. also reported a number of unresolved problems: what interests the local 'experts' represent, and how plant breeders are to interpret the results when split verdicts are returned. A rather different approach is to make a careful study over time of the choices farmers make in relation to indigenous varieties, and what it is about these varieties that appeals.

Plant breeders, it has been suggested, are not in the business of creating once-and-for-all 'ideal types'; rather, they apply their skills in pursuit of a moving target (Maurya 1989). This requires a carefully thought-out strategy for varietal selection, based on regular and continuing interaction with the farming community.

Indigenous Experimentation and Complex Innovation

A key issue in promoting complementary methods is whether it is possible to enrich existing experimental methodologies (whether practised by scientists or farmers) in order to cope with the levels of complexity normally encountered by small-scale farmers in high-risk environments.

Roland Bunch's work on strengthening existing adaptive capacities among small farmers, by providing households with the resources and stimulus to undertake additional or more elaborate experiments than hitherto, provides one example of how this can be done (Bunch, 1984). Anil Gupta (1989) stresses the other side of the coin – how agricultural scientists can learn from farmers about coping with higher levels of complexity. Drawing on experience of recent on-farm research in India and Bangladesh, he describes a number of ways in which the experiments and innovations of poor farmers can serve as a vital stimulus to creative thinking by agricultural scientists. In particular, he encourages researchers to follow up the odd, the unexpected, the absurd and the contradictory in farmers' fields. Researchers sympathetic to the idea of indigenous experimentation are sometimes most strongly attracted to those cases where farmers' and scientists' ideas appear to coincide. This may be a mistake. The most usefully thought-provoking of farmer experiments may be those that are least sensible or readily explicable in scientific terms. Even if the really odd ones prove to be bad answers, they may have started out as good questions.

Conclusion

A better understanding of farmer experimentation, and the elaboration of complementary research methods based on this understanding, has a potentially significant role to play in the search for alternatives to the so-called Green Revolution in low-resource, high-risk environments. Whether complementary methods have the capacity to foster development of complex agro-ecological innovations *in situ* – a highly attractive proposition – is yet to be determined by practical results. However, some suspect that it may be contradictory to expect to achieve such ends through a high-profile partnership of scientists and agricultural interest groups. In the long run, a vigorous defence of the intellectual autonomy of the experimenting farmer may prove to be both the better prospect and the better science.

References

Altieri, M., 'The Significance of Diversity in the Maintenance of the Sustainability of Traditional Agroecosystems', *ILEIA Newsletter*, 3 (2), 1987, pp. 3–7

Ashby, J., 'The Effects of Different Types of Farmer Participation on the Management of On-farm Trials', *Agricultural Administration*, 25, 1987, pp. 235–52

—— Quiros, C. and Rivera, Y. M., 'Farmer Participation in Technology Development: work with crop varieties', in R. Chambers, A. Pacey and L. Thrupp (eds), *Farmer First*, London, ITDG Publication, 1989

Atteh, O. D., 'Resources and Decisions: Peasant Farmer Agricultural Management and its Relevance for Rural Development Planning in Kwara State, Nigeria', PhD thesis, University of London, 1980

—— 'Nigerian Farmers' Perception of Pests and Pesticides', *Insect Sci. Applic.*, 5 (3), 1984, pp. 213–20

Biggs, S. D., 'Informal R & D', *Ceres*, 13(4), 1980, pp. 23–6

—— and Clay, E. J., 'Sources of Innovation in Agricultural Technology', *World Development*, 9, 1981, pp. 321–36

Boserup, E., *The Conditions of Agricultural Growth*, London, George Allen & Unwin, 1965

Box, L., 'Virgilio's Theorem: a method for adaptive agricultural research', in R. Chambers, A. Pacey and L. Thrupp (eds), *Farmer First*, London, ITDG Publication, 1989

Budelman, A., 'Primary Agricultural Research, Farmers Perform Field

Trials – Experiences from the Lower Tana Basin, East Kenya', *Tropical Crops Communication* (Dept of Tropical Crops, University of Wageningen), 3, 1983, pp. 10–16

Bunch, R., *Two Ears of Corn: a guide to people-centered agricultural improvement*, World Neighbors, Oklahoma, 1984

—— 'Encouraging Farmers' Experiments', in R. Chambers, A. Pacey and L. Thrupp (eds), *Farmer First*, London, ITDG Publication, 1989

Chambers, R. and Ghildyal, B. P., 'Agricultural Research for Resource-poor Farmers: the farmer-first-and-last model', *Discussion Paper No. 203*, Institute of Development Studies, University of Sussex, 1985

—— and Jiggins, J., 'Agricultural Research for Resource-poor Farmers: a parsimonious paradigm', *Discussion Paper No. 220*, Institute of Development Studies, University of Sussex, 1986

——, A. Pacey and L. Thrupp (eds), *Farmer First*, London, ITDG Publication, 1989.

Charoenwatana, T., 'Farmers and Agricultural Science', Paper for Conference on Farmers and Agricultural Research: Complementary Methods, held at the Institute of Development Studies, University of Sussex, 26–31 July 1987

Chavangi, N., 'Innovatory Participation in Programme Design: tree planting for increased fuelwood supply for rural households in Kenya', Paper for Conference on Farmers and Agricultural Research: Complementary Methods, held at the Institute of Development Studies, University of Sussex, 26–31 July 1987

Clawson, D. L., 'Harvest Security and Inter-specific Diversity in Traditional Tropical Agriculture', *Economic Botany*, 39(1), 1984, pp. 56–67

De Schippe, P., *Shifting Cultivation in Africa: the Zande System of Agriculture*, London, Routledge & Kegan Paul, 1956

Edwards, R., 'Mapping and Informal Experimentation by Farmers: agronomic monitoring of farmers' cropping systems as a form of informal farmer experimentation', Paper for Conference on Farmers and Agricultural Research: Complementary Methods, held at the Institute of Development Studies, University of Sussex, 26–31 July 1987

Farrington, J. and Martin, A., 'Farmer Participatory Research: a review of concepts and practices', *Discussion Paper 19*, Agricultural Administration (Research & Extension) Network, Overseas Development Institute, London, 1987

Fernandez, M. and Salvatiessa H., 'Participatory Technology Validation in Highland Communities of Peru', in R. Chambers, A. Pacey and L. Thrupp (eds), *Farmer First*, London, ITDG Publication, 1989

Greenland, D. J., 'Rice', *Biologist* 31(4), 1984, pp. 219–25

Gupta, A., 'Scientists' Views of Farmers' Practices in India: barriers to effective interaction', in R. Chambers, A. Pacey and L. Thrupp (eds), *Farmer First*, London, 1989

Hill, P., *The Migrant Cocoa Farmers of Southern Ghana: a study in rural capitalism*, Cambridge, Cambridge University Press, 1963

Hobsbawm, E. J. and Ranger, T. (eds), *The Invention of Tradition*, Cambridge, 1983

Igbozurike, U. M., 'Ecological Balance in Tropical Agriculture', *Geographical Review* 61, 1971, pp. 519–29

—— *Agriculture at the Crossroads: a comment on agricultural ecology*, Ile Ife, University of Ife Press, 1977

Johnny, M. M. P., 'Traditional Farmers' Perceptions of Farming and Farming Problems in the Moyamba Area', MA thesis, University of Sierra Leone, 1979

Johnson, A. W., 'Individuality and Experimentation in Traditional Agriculture', *Human Ecology* 1(2), 1972

Juma, C., 'Ecological Complexity and Agricultural Innovation: the use of indigenous genetic resources in Bungoma, Kenya', Abstract of paper for Conference on Farmers and Agricultural Research: Complementary Methods, held at the Institute of Development Studies, University of Sussex, 26–31 July 1987

Lévi-Strauss, C., *The Savage Mind*, London, Weidenfeld & Nicolson, 1966

Lightfoot, C., 'Indigenous Research and On-farm Trials', *Agricultural Administration and Extension*, 24, 1987, pp. 79–89

—— De Guia, O., Aliman, A., and Ocado, F., 'Systems Diagrams to Help Farmers Decide in On-farm Research', in R. Chambers, A. Pacey and L. Thrupp (eds), *Farmer First*, London, ITDG Publication, 1989

Maurya, D. M., 'The Innovative Approach of Indian Farmers', in R. Chambers, A. Pacey and L. Thrupp (eds), *Farmer First*, London, 1989

Millington, A. C., 'Soil Conservation Techniques for the Humid Tropics', *Appropriate Technology* 9(2), 1982, pp. 17–18

Reij, C. P., Turner, S. and Kuhlman, T., *Soil and Water Conservation in sub-Saharan Africa: issues and options*, Amsterdam: Centre for Development Co-operation Services, Free University of Amsterdam in co-operation with the International Fund for Agricultural Development (mimeo), 1986

Repulda, R., Quero, F., Ayaso, R., De Guia, O. and Lightfoot, C., 'Doing Research with Resource-poor Farmers: FSDP-EV perspectives and programs', Paper for Conference on Farmers and Agricultural Research: Complementary Methods, held at the Institute of Development Studies, University of Sussex, 26–31 July 1987

Rhoades, R. E., 'The Role of Farmers in the Creation of Agricultural Technology', in R. Chambers, A. Pacey and L. Thrupp (eds), *Farmer First*, London, ITDG Publication, 1989

—— and Booth, R. H., 'Farmer-back-to-Farmer: a model for generating acceptable agricultural technology', *Agricultural Administration* 11, 1982, pp. 127–37

—— Booth, R. H., Shaw, R. and Werge, R., 'The Role of Anthropologists in Developing Improved Technologies', *Appropriate Technology* 11(4), 1985, pp. 11–13

Richards P., *Indigenous Agricultural Revolution*, London, 1985

—— *Coping with Hunger: Hazard and experiment in an African rice farming system*, London, 1986

Rocheleau D., 'The User Perspective and the Agroforestry Action Agenda', in H. Golz (ed.), *Agroforestry*, Dordrecht, 1987

Sanghi N., 'Changes in the Organisation of Research on Dryland Agriculture', in R. Chambers, A. Pacey and L. Thrupp (eds), *Farmer First*, London, 1989

Steiner K. G., *Intercropping in Tropical Smallholder Agriculture with Special Reference to West Africa*, Eschborn, 1982

Sumberg J. and Okali C., 'Farmers, On-farm Research and New Technology', in R. Chambers, A. Pacey and L. Thrupp (eds), *Farmer First*, London, ITDG Publication, 1989

Uzozie E. 'The Changing Context of Land Use Decisions: Three Family Farms in the Yam Cultivation Zone of Eastern Nigeria' *Africa*, 51, 1981, pp. 678–93

Vaughan M. *The Story of an African Famine: Gender and famine in twentieth-century Malawi*, Cambridge, Cambridge University Press, 1987

3

Farmer-First: Achieving Sustainable Dryland Development in Africa

Robert Chambers and Camilla Toulmin

The Great Challenge of the 1990s

With agricultural development and production, on the one hand, and with poverty, on the other, the 1970s and 1980s have witnessed changes in reality and insight. By the mid-1980s, agricultural production had risen sharply in the industrial agriculture of the rich North, and in the Green Revolution (GR) agriculture of the well-watered fertile plains of the South, but not much elsewhere, in the complex, diverse and risk-prone (CDR), 'third' agriculture of the South. Food surpluses had depressed world prices, by creating a glut on the market. With the exception of Bangladesh, the most populous agricultural countries of Asia – Burma, China, India, Indonesia, Pakistan, the Philippines and Thailand – had either achieved food and foodgrain self-sufficiency or had got close to it.[1] But for much of the third, non-GR agriculture of the South, there had been a deepening crisis, with populations rising, landholdings growing smaller, environments degrading and per capita food production remaining static or declining. According to one estimate, some 1.4 billion people were dependent on CDR agriculture, with roughly 100 million in Latin America, 300 million in sub-Saharan Africa and 1 billion in Asia.[2] Short of an AIDS or similar epidemic, these were also the areas and countries where population growth rates would continue to be highest.

The problem is not one of producing enough food in the world; it is a problem of who grows it, where it is grown and who has access to it. With population growth and environmental fragility in

1. *FAO Trade Yearbook*, FAO, Rome, 1986.
2. E. C. Wolf, 'Beyond the Green Revolution: new approaches for Third World agriculture', *Worldwatch Paper 73*, Worldwatch Institute, Washington DC, 1986.

CDR areas, the problem is also one of generating sustainable livelihoods for the much larger populations of the future, enabling them to live adequately and decently where they are.[3] The alternative is that they have to migrate, often in desperation, to GR and urban areas, where they depress wages and the incomes of other poor people, or to fragile mountain, forest or semi-arid environments, where they may contribute to environmental degradation.

The great challenge for the 1990s is, then, to enable the third, CDR, agriculture to transform itself into more sustainable and productive systems, and to support many more people. To be sure, maintaining production and tackling poverty in GR areas is also vital. But the problems and solutions there are better known, although changing,[4] and receive more attention. Moreover, the normal professionalism of agricultural science has served those areas better, but fits badly with the needs and priorities of the third agriculture (Figure 3.1).

Normal Professionalism, Transfer-of-technology and the Third Agriculture

Normal professionalism means the thinking, concepts, values and methods dominant in a profession. It is usually conservative, heavily defended and reproduced through teaching, training, textbooks, professional rewards and international professional meetings. Most professional mindsets change only slowly, sometimes long after the realities and priorities have changed. This is true in the social sciences as well as in the physical and biological sciences.

In agricultural research and extension worldwide, the normal professional paradigm can be described as 'transfer-of-technology', or TOT.[5] In this model, agricultural research priorities are determined by scientists and by funding agencies. Scientists then experiment in-laboratory and on-station to generate new technology; and this is then handed over to extension for transfer to farmers. There have been many modifications and variants, but the TOT model is deeply embedded in normal professional thinking and prescription.

3. C. Conroy and M. Litvinoff (eds), *The Greening of Aid: Sustainable livelihoods in practice*, Earthscan Publication, London, 1988.
4. Wolf, 'Beyond the Green Revolution'.
5. R. Chambers and B. P. Ghildyal, 'Agricultural Research for Resource-poor Farmers: the farmer-first-and-last model', *Agricultural Administration*, 20, 1, 1985, pp. 1–30.

Figure 3.1 Three types of agriculture summarised

	Type of agriculture	Industrial	Green Revolution	Third 'CDR'
P L A C E	Main locations	Industrialised countries	Irrigated and high rainfall, high potential areas in the South	Rainfed tropics, hinterlands, most of sub-Saharan Africa, etc.
	Climatic zone	Temperate	Mainly tropical	Mainly tropical
S T A T U S	Condition	Overdeveloped	Developed	Underdeveloped
	Current production as percentage of sustainable production	Far too high	Often near the limit	Low
	Priority for production	Reduce production	Maintain production	Raise production
C H A R A C T E R	Topography usually	Flat or undulating	Flat	Undulating
	Farming system, relatively		Simple	Complex
	Environmental diversity relatively		Uniform	Diverse
	Relative stability		Low risk	High risk
	Use of external inputs	Very high	High	Low
R & D	Similarity of farmers' and research station conditions		High	Low
P R O B L E M S	Farmers consulted about research priorities	Richer farmers sometimes		Rarely
	Number of Scientists/exten-sionists per farming system		More	Fewer

CDR = Complex, diverse, risk-prone.

It is reflected in teaching, in behaviour in the field and in the rhetoric of development.

The TOT model has served industrial and GR agriculture rather well. Physical and economic conditions on research stations have been similar to those of resource-rich farms and farm families, which are typical of these two types of agriculture (Figures 3.2 and 3.3). The reductionism of normal agronomic research, in which only a few variables are manipulated, has led to simple packages suitable for uniform, controlled environments: E (the environment) has been made to fit high-yielding G (the genotype). Packages have served to standardise farming systems. The outcome has been the well-known increases in productivity per unit of land in both industrial and GR agriculture.

However, the TOT model has not done well with the third agriculture. There have been limited successes, but no great production breakthroughs comparable with the Green Revolutions with wheat, maize and rice. The explanation lies partly in the contrasts typical of CDR areas between physical and economic conditions on research stations and those of the resource-poor farms and farm families (Figures 3.2 and 3.3). It also lies in the disjuncture between the nature of CDR agriculture on the one hand, and the nature of normal professionalism on the other. This can be appreciated by examining CDR agriculture in more detail.

The complexity of any one CDR farming system has many aspects, and these also vary between farming systems. Five deserve mention. First, physically, CDR farm-holdings often comprise sloping lands with a variety of conditions of soil, slope, shade, aspect and water supply, and sometimes include lands in different ecological zones on the same holding, and with energy and nutrient linkages with common property resources (such as where livestock grazed on common grazing lands during the day are used to manure farmers' fields each night). Second, in their internal linkages, CDR farming systems typically involve and rely on complex interactions between crops, livestock, grasses, trees, and sometimes fish and insects. Intercropping and agro-forestry in their many forms are typical of this sort of complexity. Third, CDR farming systems are complex temporally, with many different processes and activities at different times of the year. Fourth, CDR farming systems entail several or many enterprises, often off-farm as well as on-farm; many species of useful plants and animals are husbanded, and often these are multi-purpose and multi-product. Finally, compounding all these complexities, CDR farming systems are particularly intimately interlinked with the farm house-

Figure 3.2 Typical contrasts in physical conditions

	Research/ experiment station	Resource- rich farm (RRF)	Resource- poor farm (RPF)
Soils	Deep fertile few constraints	Few effective constraints	Shallow, infertile, often severe constraints
Macro- and micro-nutrient deficiency	Rare, remediable	Occasional	Quite common
Plot size and nature	Large, square	Large	Small, irregular
Hazards	Nil or few	Few, usually controllable	More common – floods, droughts, animals grazing crops, etc.
Irrigation	Usually available	Usually available	Often non-existent
Size of management unit	Large, contiguous	Large or medium, contiguous	Small, often scattered and fragmented
Natural vegetation	Eliminated	Eliminated or highly controlled	Used or con-trolled at micro-level

Source: Chambers and Jiggins (1986), adapted from Chambers and Ghildyal (1985: 6).

hold, its labour power, social structure and economy, given the frequent absence of effective markets for many farm inputs and credit.

In addition, CDR agriculture often presents diversity of farming systems within short distances, corresponding with differences which are ecological, social and economic, for example, in accessibility to markets. It is also often risky, being usually rainfed and subject to the vagaries of climate, without the stabilising effects of reliable irrigation.

Normal agricultural science does not fit well with these characteristics. The complexity of CDR agriculture presents interactions that are difficult for scientists to manage and study. Some lie in the

Figure 3.3 Typical contrasts in social and economic conditions

	Research experiment station	RRF family	RPF family
Access to seeds, fertilisers, pesticides and other purchased inputs	Unlimited, reliable	High, reliable	Low, unreliable
Source of seeds	Foundation stocks, and breeders' seed high quality	Purchased, high quality	Own seeds
Access to credit when needed	Unlimited	Good access	Poor access and seasonal shortages of cash when most needed
Irrigation, where facilities exist	Fully controlled by research station	Controlled by farmers or by others on whom s/he can rely	Controlled by others, less reliable
Labour	Unlimited, no constraint	Hired, few constraints	Family, constraining at seasonal peaks
Prices	Irrelevant	Lower than RPF for inputs. Higher than RPF for outputs	Higher than RRF for inputs. Lower than RRF for outputs
Priority for food production	Neutral	Low	High
Access to extension services	Good but one-sided	Good, almost all material designed for this category	Poor access; little relevant material

Source: Chambers and Jiggins (1986), adapted from Chambers and Ghildyal (1985: 6).

gaps between dominant disciplines (concerning agro-forestry, tree fodders, crop residues, biological energy use, etc.): normal science homes in on its primary concern – crops for agronomists, livestock for animal scientists, trees for foresters – rather than their linkages. Some opportunities lie in complex simultaneous innovation, where several factors must be changed at the same time, as with harvesting soils, nutrients and water, or introducing a cover crop to inhibit weed growth, or much agro-forestry where there are tree–crop, tree–livestock or tree–crop–livestock interactions. For scientists tied to respectable statistical methods, these complexities can be an unmanageable nightmare: for if they simplify them until they are measurable, they destroy the complexities which are their strength.

Precisely this bad fit of CDR agriculture with normal professionalism has served to conceal its potential. When the simple packages generated in the TOT mode are not adopted in CDR areas, the conclusion can easily be drawn that the areas themselves lack potential. So they are often referred to as 'resource-poor' or 'low-resource' areas. But a case can be made out that their sustainable potential as a multiple of present performance, is considerable[6] and may be far greater than sustainable potential as a multiple of present performance for GR agriculture. The latter is, in some parts of the developing world, already near its limit, and is itself tending towards more diversified patterns of cropping.

This misfit has been compounded by diversity and by scientists' motivation. For scientists, there is a problem of cost-effectiveness and professional rewards. Any innovation, such as a new variety or new practice, is likely to fit conditions and needs of far fewer farm families in CDR areas than in GR areas, which are or can be made so much more uniform. This makes work harder to justify economically, and also reduces the prestige and incentives of the work for scientists looking for the big breakthroughs. This difficulty is compounded by the presence of far fewer scientists per farming system.[7] This reflects the past unpopularity of CDR agriculture, and its low status and low political priority. Understandably,

6. R. Bunch, 'Case Study of Guinope Integrated Development Programme, Guinope, Honduras', Paper for Only One Earth: Conference on Sustainable Development, organised by the International Institute for Environment and Development, London, 28–30 April 1987; R. Chambers, 'Sustainable Rural Livelihoods: a strategy for people, environment and development', *Commissioned Study 7*, Institute of Development Studies, University of Sussex, 1987.

7. R. Chambers and J. Jiggins, 'Agricultural Research for Resource-poor Farmers: a parsimonious paradigm', *IDS Discussion Paper 220*, Institute of Development Studies, University of Sussex, 1986.

irrigated Green Revolution agriculture has been preferred by scientists and PhD students for reasons including accessibility, ease of control, and predictability of experiments, research papers and PhDs.

To gain increased attention for the problems faced by CDR agriculture and the promise of alternative research approaches requires a shift not only in the structure of incentives for scientists, but also within research and extension systems. The latter would in most cases require quite major changes in how extension staff are assessed and rewarded, and the overall ethos surrounding extension–farmer relations.

For the third, CDR agriculture, the TOT paradigm is in crisis. At the extreme, the research priorities and locations are wrong, the messages do not fit, the packages are rejected, and the bad experience is attributed either to farmers' ignorance (prescription – more and better extension), or to farm-level constraints (prescription – identify and ease the farm-level constraints and simplify and control the farm to make it more like the research station). This approach frequently brings failure, as it seeks to reduce just those elements of CDR farming systems which provide its strengths – its complexity and adaptiveness. Farmers' own research methods are usually very different and aim to experiment with new techniques and varieties on a small scale, to assess their value within the constraints imposed by local conditions.

The crisis is also one of direction. Often, CDR farmers reduce their risks by making their farming systems more complex. In terms of agro-ecology, this is analogous to the greater resilience in face of risk associated with complex compared with simple ecosystems. Normal TOT seeks to simplify, and thereby increases vulnerability to risk, and emphasises purchased inputs which, for CDR farmers, often introduce problems of reliable access. For their part, CDR farm families tend to diversify (both to increase benefits from production and to spread risks) and to rely on factors of production that are under their control.

Farmer–first: The Complementary Paradigm

The crisis has led to questioning the very processes that generate agricultural technology, and to the exploration of new approaches. Increasingly during the 1980s, innovators in the agricultural and social sciences have been working with CDR farmers to find solutions to these problems. By concentrating on what they find to work, they have evolved a new paradigm for agricultural research and extension. The approaches of this paradigm have been given various labels: farmer-back-to-farmer;[8] farmer-first-and-last;[9] farmer participatory research;[10] recherche-développement;[11] and approach development.[12]

The name does not much matter, but farmer participation is one key element. For inclusiveness and brevity, we shall try to capture the essence of these approaches with the title farmer-first (FF). It should be noted here, however, that we need to see the farmer within the context of the household and community of which he or she is part (cf. Hunt's full household economy model in this volume). We would also want to include within the term 'farmer' members of the pastoral community who have been particularly neglected by the research and extension services to date.

The essence of FF is that it reverses some parts of the TOT process, which have tended to go unquestioned.[13] A reversal of explanation looks for reasons why farmers do not adopt new technology in deficiencies in the technology and the process which generated it, rather than in farmers' ignorance. A reversal of learning has researchers and extension workers learning from farmers. Location and roles are also reversed, with farms and farmers central, instead of research stations, laboratories and scientists.

In this framework, much farming systems research can be seen as an extension of TOT: information has been obtained from farmers by outsiders, and analysed by them to decide what would be good for the farmers, leading to the design of experiments for testing and adaptation. In contrast, FF reverses roles. Analysis, choice and experimentation are conducted by and with farmers themselves, with outsider professionals in a facilitating and support role.[14]

8. R. E. Rhoades and R. H. Booth, 'Farmer-back-to-farmer: a model for generating acceptable agricultural technology', *Agricultural Administration*, vol. 11, 1982, pp. 127–37.
9. Chambers and Ghildyal, 'Agricultural Research for Resource-poor Farmers'.
10. J. Farrington and A. Martin, 'Farmer Participatory Research: a review of concepts and practices', *Discussion Paper 19*, Agricultural Administration (Research and Extension) Network, Overseas Development Institute, London, June 1987.
11. *Les Cahiers de la Recherche-Développement*, no. 1, June 1983.
12. U. Scheuermeier, 'Approach Development: a contribution to participatory development of techniques based on a practical experience in Tinau Watershed Project', Nepal, LBL, Landwirtschaftliche Beratungszentrale, Lindau, Switzerland.
13. Conroy and Litvinoff, *The Greening of Aid*.
14. R. Chambers, 'Sustainable Rural Livelihoods: a key strategy for people, environment and development, in ibid., pp. 1–17; P. Gubbels, 'Peasant Farmer

There are now many published sources of FF experience. They include *Experimental Agriculture*[15] with selected papers from the workshop on Farmers and Agricultural Research: Complementary Methods, held at the Institute of Development Studies, Sussex University, the Administration (Research and Extension) Network of the Overseas Development Institute, London and in particular 'Farmer Participatory Research: A Review of Concepts and Practices',[16] and papers from the Workshop on Participatory Technology Development, held in April 1988 by the ILEIA, in the Netherlands.

Accessible examples of FF experience worldwide include the work of Jacqueline Ashby and her colleagues at CIAT in Colombia,[17] of Roland Bunch and World Neighbors,[18] of D. M. Maurya in India,[19] the pioneering rapid rural appraisal (RRA) work of the University of Khon Kaen,[20] and RRA Notes.[21]

In Africa, examples of FF experience are fewer and more recent. Here, it has usually been the non-governmental organisations (NGOs) which have carried out the most innovative work in this field. Limited resources within national research and extension systems and the hierarchical structure of the latter have both constrained the extent to which staff have been either able or willing to maintain much contact with farmers, especially those in more distant, marginal farming areas.[22]

Agricultural Self-development', in *Participatory Technology Development*, ILEIA Newsletter, vol. 4, no. 3, 1988, pp. 11–14.

15. J. Farrington (ed.), *Experimental Agriculture*, vol. 24, part 3, 1988.

16. Farrington and Martin, 'Farmer Participatory Research'.

17. J. Ashby, C. Quiros and Y. Riviera, 'Farmer Participation in On-farm Varietal Trials', *Discussion Paper 22*, Agricultural Administration (Research and Extension) Network, Overseas Development Institute, London, December 1987.

18. R. Bunch, *Two Ears of Corn: a guide to people-centered agricultural improvement*, World Neighbors, Oklahoma, 1985.

19. D. M. Maurya, A. Bottrall and J. Farrington, 'Improved Livelihoods, Genetic Diversity and Farmer Participation: a strategy for rice breeding in rainfed areas of India', *Experimental Agriculture*, 24, part 3, 1988, pp. 311–20.

20. Khon Kaen University, Proceedings of the 1985 International Conference on Rapid Rural Appraisal, Rural Systems Research and Farming Systems Research Project, Khon Kaen, Thailand, 1987; G. Lovelace, S. Subhadira and S. Simaraks (eds), *Rapid Rural Appraisal in Northeast Thailand: Case studies*, KKU-Ford Rural Systems Research Project, Khon Kaen University, Thailand, 1988.

21. *RRA Notes* no. 6, The Sustainable Agriculture Programme, International Institute for Environment and Development, London.

22. Farrington and Martin, 'Farmer Participatory Research'.

Farmer-first Work in Africa

Work within Africa with FF methods is mainly confined to the experience of NGOs, who have needed to adopt a learning-process approach to their development programmes. It is important to recognise that FF methods are not seen as substitutes for all other forms of enquiry. An essential first step will always be the gathering of useful information already existing on a particular region. The cases below concern work with FF methods by researchers and organisations that were already familiar with the institutional and technical constraints facing the communities with which they were working.

Oxfam's agro-forestry project in Yatenga region (Burkina Faso) is a notable example of research and experimentation by an NGO and the local community in developing manageable techniques for soil and water conservation.[23] Having begun, as its name suggests, as a project aimed at tree-planting activities, its direction changed greatly to focus on what farmers felt was much more important – the question of maintaining crop-yields on their farmland. Simple yet effective techniques have been developed to conserve soil and water better, by the use of low, rock bunds built along field contours. These bunds help slow down the flow of water across the field and thus trap water and soil within the field. Crop yields are estimated to be from 20 to 50 per cent higher on treated versus untreated land, with the stabilisation of yields in years of low rainfall particularly important for farmers.

In Zimbabwe, the Zvishavane Water Resources Project, an indigenous NGO, has been involved in a process of farmer-based research, focused on developing strategies for improving the productivity and sustainable use of small natural wetland patches in this otherwise semi-arid region.[24] The crucial role of these wetland patches was identified by farmers through a process of discussion about natural resource availability and use within the area. The greater level of fertility and soil moisture allows for regular cropping of these patches, with yields less vulnerable to rainfall variability. The example of one particular farmer who has developed a highly diverse way of using his wetland, with crops, trees and

23. P. Harrison, *The Greening of Africa*, Paladin, London, 1987, pp. 166–9; R. M. Rochette (ed.), *Le Sahel en lutte contre la desertification*, GTZ/Margraf, Weikersheim, 1989, pp. 221–38.
24. P. Maseko, I. Scoones and K. Wilson, 'Farmer-based Research and Extension', *Enhancing Dryland Agriculture*, ILEIA Newsletter, vol. 4, no. 4, 1988, pp. 18–19.

ponds, receives visiting farmers. These visits are then the occasion for much debate between the farmers as to the options they can pursue. It is clear from the case study that farmers often have their own ideas of what might be good development options, but they may need some support in experimenting with these.

In Senegal, the locally based NGO Environment et Dévéloppement du Tiers Monde (ENDA) has been supporting a programme of 'participative training' for students at the Institut National du Dévéloppement Rural at Thie, inland from Dakar.[25] This programme has taken as its focus local farmers' perception that yields and the fertility of soils have been declining. Students are brought face to face with the conditions under which farmers currently operate, and are introduced to an approach that emphasises joint reflection, research, debate and experimentation. The initial workshop comprised six stages: (1) reading the landscape; (2) listening to farmers describe the history of their plots of land; (3) noting down methods of fertilising soils and the results gained; (4) sampling and analysis of soils from certain plots; (5) evaluation of results and comparison across the sample of villages taken; and (6) presentation of the results to farmers and subsequent discussion. The initial workshop is seen only as a starting point for a longer process of joint research, having identified key subjects of importance. For the workshop described, the main problem areas that emerged included: problems in regenerating and protecting *Acacia albida*, a leguminous tree of value within agro-forestry systems; reduced access to animal dung, due to the fall in regular manuring contracts between farmers and Fulani pastoralists, as a result in part of the disappearance of fallow and consequent falling pasture availability; and the erosion of soils by wind and water. Having defined problem areas, the workshop went on to suggest measures to address these, such as trying to rebuild livestock–crop linkages, making better use of existing organic matter by composting, and construction of low bunds on field contours to slow sheet–erosion of soils. These measures then provide the focus for longer-term collaboration between the farmers and students. Good visual aids are important in generating debate during the last stage of the workshop; several farmers commented on how looking at a photograph of their own field has enabled them to see more clearly how bare and denuded is their land.

Approaches to community development using FF techniques

<hr>

25. ENDA-GRAF, 'Pour une recherche-formation action sur la fertilité des sols: une étude de cas en milieu sahélien', *ENDA document de base*, no. 270, ENDA Tiers Monde, Dakar, 1987.

have been developed by the Ethiopian Red Cross Society for their programme in Wollo Region.[26] In an initial exercise in March 1988, staff from the ERC and from IIED, London and IDS, Sussex carried out a survey of two villages in Wollo. This tested the applicability of RRA/FF techniques to the development work of the ERCS and sought possible innovations in the two villages studied that might bring equitable and sustainable progress for these communities. An eight-day workshop brought together staff from a number of organisations to assess the main constraints and opportunities present in each locality. Through a discussion process between workshop members and villagers, the priority needs of different villagers were identified. The investigation phase was followed by an assessment of possible 'best bets', ranging from provision of credit, to reforestation and small-scale irrigation development. Each 'best bet' was then examined in terms of its likely impact on a number of criteria, such as productivity, stability of income produced, equitability, sustainability and cost. The interventions suggested by the workshop now form part of the ERCS programme in Wollo.

In Kenya, the National Environment Secretariat has been using participatory research techniques to help develop resource management plans at village level in Machakos District.[27] Bringing together researchers, local and national government officers, leaders of local women's groups and village people, workshops have been held to discuss a series of possible development options. These meetings have thrown up a large number of ideas about the priorities placed by different groups on such subjects as water, human health, tree planting and marketing. However, it was not always easy to get women to participate fully in the discussions, even when these dealt with issues of greatest interest to them. Visual materials again played a crucial role in providing a focus for discussion, particularly where different languages were being used. By bringing together local government officers from the different technical services, the workshop allowed them to discuss problems common to their separate sectors in a way that does not normally happen, due to sectoral allocation of responsibilities.

World Neighbors in Mali have been building up a network of farmer-researchers with whom they test and identify promising

26. Ethiopian Red Cross Society, *Rapid Rural Appraisal: A closer look at rural life in Wollo*, ERCS, Addis Ababa and IIED, London, 1988.
27. C. Kabutha and R. Ford, 'Using RRA to Formulate a Village Resources Management Plan, Mbusanyi, Kenya', *RRA Notes 2*, International Institute for Environment and Development, London, 1988, pp. 4–11.

seed varieties.[28] The programme started with diagnosis by peasant farmers of agricultural problems by focusing on debate around three main questions:

1 How has farming changed since the time of your father and grandfather?
2 What are the main problems you face as a farmer, and what are their causes?
3 How have you tried to cope with these difficulties and with what success?

These discussions throw up much information about farmers' own perceptions of change and their coping strategies. Following the discussion of options for future development, villagers are helped to visit research stations or other communities where the benefits from different options can be investigated. Farmers near Segou in central Mali were particularly concerned to find a reliable short-cycle millet variety that also satisfied their criteria regarding taste and storage. Trials were set up by farmers to examine the performance of different imported seed varieties, establishing control plots next to their own fields and spanning a range of agro-ecological conditions. Following the harvest, farmers from the four participating villages were brought together to evaluate the performance of each variety and to make recommendations about appropriate techniques (such as date of sowing, soil type, etc.). Following this assessment there has been a rapid and widespread diffusion of these seed varieties, entirely managed by villagers themselves.

In the latter half of the 1980s, FF methods evolved quickly. Many forms and variants now are being tried. Some of the contrasts with TOT are presented in Figure 3.4. While not all of these are found all the time, and some can be followed without others, they are mutually reinforcing and cohere as a paradigm contrasting with and complementary to TOT. While farmer participation is a widespread and crucial element, FF goes beyond that to influence decisions and methods which may not involve farmers directly and immediately , for example concerning on-station research.

28. Gubbels, 'Peasant Farmer Agricultural Self-development'.

Figure 3.4 Transfer-of-technology and farmer-first compared

	TOT	FF
Main objective	Transfer technology	Empower farmers
Analysis of needs and priorities by	Outsiders	Farmers assisted by outsiders
Transferred by outsiders to farmers	Precepts	Principles
	Messages	Methods
	Package of practices	Basket of choices
The 'menu'	Fixed	*A la carte*
Farmers' behaviour	Hear messages	Use methods
	Act on precepts	Apply principles
	Adopt, adapt or reject package	Choose from basket and experiment
Outsiders' desired outcomes	Widespread adoption of package	Wider choices for farmers
emphasise		Farmers' enhanced adaptability
Main mode of extension	Agent-to-farmer	Farmer-to-farmer
Roles of extension agent	Teacher	Facilitator
	Trainer	Searcher for and provider of choice

Farmers' Analysis, Choice and Experimentation

One sequence which recurs in farmer participatory research and extension activities is an iterative process of farmers' analysis, choice, and experiment (FACE) followed by evaluation and extension. The main activities of farmers and roles of outsiders are:

Farmers' activities	*New roles for outsiders*
● Analysis	Convenor, catalyst, adviser
● Choice	Searcher and supplier
● Experiment	Supporter and consultant

The actors and activities are presented diagrammatically in Figure 3.5.

Let us consider the main activities in turn.

Analysis

Farmers' analysis can be promoted and supported in many ways:

1 Sequences of farmers' group discussions and visits.[29]
2 Inspection and discussion – visiting other farmers, research stations, or trial sites.[30]
3 Innovator workshops, where farmer innovators meet and discuss and compare their new practices.[31]
4 The use of key priming questions by outsiders, such as: 'What would an ideal variety look like to you?' 'What would you like your landscape to look like in the future?' 'What do you farmers talk about when you get together?' 'Why do other farmers have different practices to you?' And the unhurried sequence: 'What was farming like when you were young, how has it changed, what problems have you faced, with what have you tried to tackle them, and with what results?'[32]
5 Visual aids to analysis such as seasonal diagramming,[33] aerial photographs and overlays, systems diagramming[34] and charts

29. J. Sumberg and C. Okali, 'Farmers, On-farm Research and the Development of New Technology', *Experimental Agriculture*, 24, part 3, 1988, pp. 321–31; Lovelace et al., *Rapid Rural Appraisal in Northeast Thailand*; L. Tung and F. T. Balina, Proceedings, Philippines Upland Research and Extension Training Workshop, ATI, NTC-Visayas, Philippines, 19–24 June 1988.

30. Ashby et al., 'Farmer Participation in On-farm Varietal Trials'; Gubbels, 'Peasant Farmer Agricultural Self-development'.

31. Z. Abedin and F. Haque, 'Learning from Farmer Innovation and Innovator Workshops; experiences from Bangladesh', Paper for the Conference on Farmers and Agricultural Research: Complementary Methods, Institute for Development Studies, University of Sussex, 26–31 July 1987; Ashby et al., 'Farmer Participation in On-farm Varietal Trials'.

32. ENDA-GRAF, 'Pour une recherche-formation action sur la fertilite des sols'; *Pour une pédagogie de l'autopromotion. Nouvelle edition pour les animateurs villageois.* Groupe de Recherche et d'Appui pour l'Autopromotion Paysanne (GRAAP), Karthala/GRAAP, 1985. *Nouvelles paroles de brousse*, Karthala, Paris, 1988, GRAAP, BP 785, Bobo-Dioulasso, Burkina Faso; La lettre du Reseau Recherche-Dévéloppement, Dossier-Gestion de terroirs, GRET, Paris, no. 10, March 1989.

33. G. Conway, 'Diagrams for Farmers', Paper for the Conference on Farmers and Agricultural Research: Complementary Methods, Institute of Development Studies, University of Sussex, 26–31 July 1988; Ethiopian Red Cross Society, *Rapid Rural Appraisal*.

34. Tung and Balina, Proceedings, Philippines Uplands and Research and Exten-

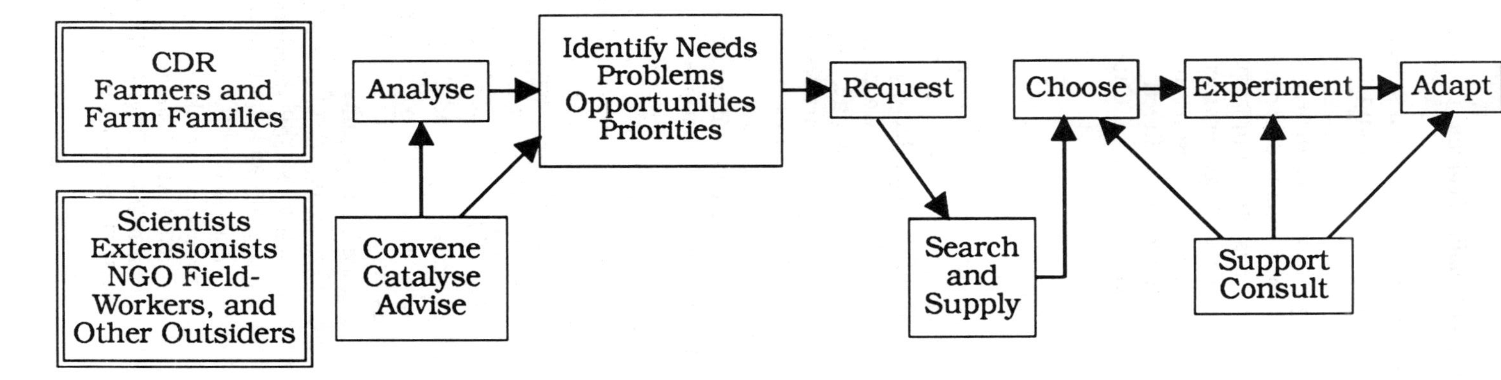

Figure 3.5 Activities in the farmers' analysis-choice-experiment (FACE) approach

representing farmers' information systematised by outsiders,[35] drawn on boards or on the ground.

Methodological questions are many, and much remains to be learnt. Analysis can raise many different sorts of issue. In CDR areas, security of tenure is often a prerequisite for farmers taking a long view. Or relations with government departments may turn out to be crucial. It may be necessary to tackle priorities such as these before those that are more directly agricultural. Or analysis may lead straight to experimentation. Often, though, it will lead to search.

Search

Participatory analysis often generates demands for information and material. CDR farmers want and need wide choice and enhanced adaptability. The role of the outsider, whether researcher or extension agent, is to look for and supply a range of information about practices and potentials, and a range of genetic material. The demand is not for the package of practices for normal research and extension, but for a basket of choices.

Methodological questions refer especially to the organisation of extension and research. Extension information systems have to be stood on their head, passing requests up first, before messages down.

Choice

There are methodological questions about how best to elicit and support farmers' criteria and choices. This may be done, for example, by providing mini-kits containing several varieties of a crop, and several fertilisers, for farmers to test and choose from on their own. It is also important to keep in mind the wide range of criteria important to different members of a community: choices made by men and women, rich and poor, are likely to differ and to reflect their particular needs and constraints. One example – of group discussion – illustrates how this may be done in order to explore in some detail the characteristics of different tree species and the various uses to which each species may be put.

---

sion Training Workshop.

35. Kabutha and Ford, 'Using RRA to Formulate a Village Resources Management Plan'.

In a workshop in a group of villages around Khartoum (Sudan), which examined incentives for tree management at individual, group and regional level, one interview was held with a group of migrants from western Sudan and another with local settled farmers. Each group was asked to name the six most important trees to them, and the names of these trees were then written down on separate pieces of paper. Each group was then asked, for each pair of trees, which one they preferred and the reasons for their choice, from which a ranking of the least to the most preferred tree was derived. This procedure brought out a large amount of information about the needs that different trees satisfy – some functional (wood for building, leaves for fodder), some aesthetic (ornamental value and density of shade) and some cultural (value in funeral and marriage ceremonies). But, in particular, it showed the marked difference in criteria used and trees valued between the two groups interviewed, and the consequent need to consider carefully the different groups present within a community when investigating people's choices.[36]

Experimentation

Finally, farmers themselves experiment, and adapt technology.[37] Here what is often most important is to transfer to them not packages and precepts, but principles and methods. A famous example of the transfer of a principle is the International Potato Center's experience with diffused light storage in potatoes. Farmers themselves discovered that sprouting in storage – a problem with new varieties – was inhibited by diffused light storage. Scientists learnt from the farmers, and transferred the principle internationally. But there was no standard store to be built; farm families did not adopt a design but applied a principle, in a myriad of locally adapted different ways. An example of the transfer of a method is provided by World Neighbors, who have a simple procedure for enabling farmers to conduct their own trials more systematically.[38]

36. I. Scoones and J. Pretty, 'Rapid Rural Appraisal for Economics: exploring incentives to tree management in Sudan', International Institute for Environment and Development, London.

37. A. W. Johnson, 'Individuality and Experimentation in Traditional Agriculture', *Human Ecology*, vol. 1, no. 2, 1972, pp. 149–59; P. Richards, *Indigenous Agricultural Revolution: Ecology and Food Production in West Africa*, London, 1985; R. Rhoades, 'Farmers and Experimentation', *Discussion Paper 21*, Agricultural Administration (Research and Extension) Network, Overseas Development Institute, London, December 1987.

38. R. Bunch, 'Small Farmer Research: the key element of permanent agricultural improvement', Paper for the Conference on Farmers and Agricultural Research:

Many methodological questions remain. One persistent problem is allowing and enabling farmers to 'own' their experiments and not to be dominated by outsiders. Enhancing farmers' capacity to experiment remains a major frontier on which much progress is needed and can be expected. As Farrington and Martin note,[39] there are very few cases in which researchers and farmers have worked side-by-side in the design and conduct of *formal* trials. Loss of statistical accuracy and of full control methods are frequently cited as posing the main problems for such collaboration.[40] However, as discussed below, evaluation of different practices does not necessarily require formal statistical analysis.

Evaluation and Extension

In the FF mode, evaluation is not by scientists' peers but by farmer adoption. For D. M. Maurya (personal communication) whether a line justifies the bulking of seed depends on whether the farmers who try it are asked for seed by other farmers. With farmers' inspections of one anothers' fields and trials, evaluation and extension merge. Extension is not top-down, as often in the T and V mode in practice, but lateral, from farmer to farmer, as with peanuts after rice in north-east Thailand,[41] with soil erosion control in the Philippines (S. Fujisaka: personal communication), and in the approach of World Neighbors.[42] NGOs in Sahelian Africa have recognised the potential benefits of farmers learning from other farmers, and several run exchange visits between project areas. These have brought, for example, Malian farmers to see soil and water conservation work in Burkina Faso, and pastoral herders from Chad to see work in Senegal.

The FF paradigm is still evolving and will never have a final shape, since it is organic rather than a structure. All the same, there are recurring elements which hang together and support each other. One is the resonance between enhancing the adaptability of

---

Complementary Methods, Institute of Development Studies, University of Sussex, 26–31 July 1987; Gubbels, 'Peasant Farmer Agricultural Self-development'.

39. Farrington and Martin, 'Farmer Participatory Research'.

40. Ibid.; J. McCracken, *Participatory Rapid Rural Appraisal in Gujarat: a trial model for the Agha Khan Rural Support Programme (India)*, International Institute for Environment and Development, London, 1988.

41. A. Jintrawet, S. Smutkupt, C. Wongsamun, R. Katawetin and V. Kerdsuk, 'Extension Activities for Peanuts after Rice in Ban Sum Jan, Northeast Thailand: a case study in farmer-to-farmer extension methodology', Khon Kaen: Farming Systems Research Project, Khon Kaen University, Thailand, June 1985.

42. Bunch, *Two Ears of Corn*.

farmers through widening their choice and knowledge, and enhancing the adaptability of outsiders – scientists, extensionists and NGO staff – through widening theirs. For farmers the choices are of practices and plants; for outsiders, of approaches and methods. For farmers, the adaptability is to uncertain climatic and economic conditions; for outsiders, it is to needs, opportunities and insights as they arise. For all, decentralisation and reversals of authority to those 'below' are entailed: to empower farmers to analyse, choose, experiment and evaluate; and to empower outsiders, however junior, to use their initiative and choose their methods to fit local conditions. It is important to recognise that FF methods imply a radical shift in the power relations existing between experts/ professionals, and local people. If such a shift is to be effective it may well produce demands for change in other power relationships affecting villagers' lives.

FF thus has its own style, which is decentralised and democratic, in which there is mutual respect and service between outsiders and farmers. Personality is here a key variable. FACE may not be a bad acronym, since the quality of the face-to-face interactions of farmers and outsiders are crucial. A personal impression is that those who have succeeded in pioneering FF approaches have been sympathetic people, who empathise with farmers and respect and like them. This cannot be expected of all outsiders, but the fascination and psychic rewards of working closely with farmers and learning from and with them are so high, that more and more outsiders may be attracted to this mode.

Reflections for the Future

The argument for the FF paradigm to complement TOT has been developed here in terms of the third, CDR, agriculture, but its application is not necessarily so limited. It may increasingly fit the trends in GR agriculture towards complexity and diversity, particularly where the withdrawal or reduction of input subsidies in both GR and industrial agriculture permits and encourages on-farm diversification towards complexity. FF approaches and methods, devised and evolved to meet the special challenges of CDR agriculture, may in the 1990s be found to apply more and more in GR and industrial agriculture, helping the 1990s to become a decade, worldwide, of diversification.

For the present, though, the higher priority appears to lie in CDR agriculture, evolving and testing methods, and striving for cost–effectiveness, spread and sustainability. This raises many questions, including these:

1 To what extent, and how, can the FF paradigm be parsimonious, that is, sparing in its demands on outsiders' time so that many more of the diverse farming systems can be served? How can FF approaches be given greater coherence to help spread knowledge of these techniques without stifling the vigour and variety of thinking from which such tools are being developed?

2 How can FF approaches and methods be assessed and evaluated, to identify what works, and how well it works and in what conditions?

3 How can FF pioneers in national and international agricultural research systems, and in national extension systems, be encouraged, supported and rewarded, in a sustained manner, with freedom to behave in new ways?

4 How can practitioners learn efficiently from their experience and pass it on to others?

5 How can new syllabuses, textbooks and training courses be evolved to include FF experience and methods?

6 How can collaboration between the NGO, research and extension communities be encouraged, enabling them to build on their particular skills and experience?

7 Within the Sahelian context, how can FF approaches meet the challenge posed by the pastoral livestock sector, for which past development efforts have been especially unsuccessful?

8 How can the FF paradigm support and affirm the policy shift (in rhetoric, at least) of many governments in sub-Saharan Africa towards greater decision-making and control by local communities over their own resources and development plans? What potential is promised by participatory rural appraisal, as pioneered in Kenya[43] and Gujarat?[44]

9 Where communities become closely involved in diagnosis of their problems and constraints, how can this analysis be backed up effectively? There are dangers from unleashing expectations amongst a community which cannot subsequently be met.

10 Given the heterogeneity of rural communities, in terms of wealth, gender, political power, age, skills, etc. how can FF practitioners take account of this diversity in their methods?

43. Kabutha and Ford, 'Using RRA to Formulate a Village Resources Management Plan'.

44. McCracken, *Participatory Rapid Rural Appraisal in Gujarat*.

For example, women may not have time to attend workshops or lengthy discussions due to their many work commitments and care of children.

It is too early to say what the ultimate potential of FF approaches and methods will be. It is not too early to say that finding out that potential is a priority; for on it may depend the sustainable livelihoods of many millions of the poorest in the 1990s and the twenty-first century.

Acknowledgements

Many people have provided valuable comments on this paper. We would particularly like to thank Diana Hunt, Simon Maxwell, John Farrington and Janice Jiggins, who have made extensive comments on previous drafts of this paper.

References

Abedin, Z. and Haque, F., 'Learning from Farmer Innovation and Innovator Workshops: experiences from Bangladesh', Paper for the Conference on Farmers and Agricultural Research: Complementary Methods, Institute of Development Studies, University of Sussex, 26–31 July 1987

Ashby, J., Quiros, C. and Rivieria, Y., 'Farmer Participation in On-farm Varietal Trials', *Discussion Paper 22*, Agricultural Administration (Research and Extension) Network, London, Overseas Development Institute, December 1987

Baker, G., Knipscheer, H. and Nieto, J. de Souza, 'The Impact of Regular Research Field Hearings (RRFH) in On-farm Trials in Northeast Brazil', *Experimental Agriculture*, 24, part 3, 1988, pp. 281–8

Bunch, R., *Two Ears of Corn: a guide to people-centered agricultural improvement*, World Neighbors, Oklahoma, 1985

—— 'Case Study of Guinope Integrated Development Programme, Guinope, Honduras', Paper for Only One Earth: Conference on Sustainable Development, organised by the International Institute for Environment and Development, London, 28–30 April 1987

—— 'Small Farmer Research: the key element of permanent agricultural improvement', Paper for the Conference on Farmers and Agricultural Research: Complementary Methods, Institute of Development Studies, University of Sussex, 26–31 July 1987

Byerlee, D., 'Maintaining the Momentum in post-Green Revolution Agriculture: a micro-level perspective from Asia', *MSU International Development Paper No. 10*, Michigan State University, East Lansing, 1987

Cahiers de la Recherche-Développement, no. 1, June 1983

Chambers, R., 'Normal Professionalism, New Paradigms and Development', *Discussion Paper 227*, Institute of Development Studies, University of Sussex, December 1986

—— 'Sustainable Rural Livelihoods: a strategy for people, environment and development', *Commissioned Study* 7, Institute of Development Studies, University of Sussex, 1987

—— 'Sustainable Rural Livelihoods: a key strategy for people, environment and development', in Conroy and Litvinoff (1988), pp. 1–17

—— and Ghildyal, B. P., 'Agricultural Research for Resource-poor Farmers: the farmer-first-and-last model', *Agricultural Administration*, 20, 1, 1985, pp. 1–30

—— and Jiggins, J., 'Agricultural Research for Resource-poor Farmers: a parsimonious paradigm', *IDS Discussion Paper 220*, Institute of Development Studies, University of Sussex, 1986

Conroy, C. and Litvinoff, M. (eds), *The Greening of Aid: Sustainable Livelihoods in Practice*, London, Earthscan Publications, 1988

Conway, G., 'Diagrams for Farmers', Paper for the Conference on Farmers and Agricultural Research: Complementary Methods, Institute of Development Studies, University of Sussex, 26–31 July 1987

ENDA-GRAF, 'Pour une recherche-formation action sur la fertilité des sols: une étude de cas en milieu sahélien', *ENDA document de base*, no. 270, ENDA Tiers Monde, Dakar, 1987

Ethiopian Red Cross Society, *Rapid Rural Appraisal: A closer look at rural life in Wollo*, ERCS, Addis Ababa and IIED, London, 1988

FAO Trade Yearbook, Rome, FAO, 1986

FARMIIS, *Farm and Resource Management Institute Information Service Newsletter*, Farm and Resource Management Institute, Philippines, December 1987

Farrington, J. (ed.), *Experimental Agriculture*, vol. 24, part 3, 1988

—— and Martin, A., 'Farmer Participatory Research: a review of concepts and practices', *Discussion Paper 19*, Agricultural Administration (Research and Extension) Network, Overseas Development Institute, London, June 1987

Groupe de Recherche et d'Appui pour l'Autopromotion Paysanne (GRAAP), *Pour une pédagogie de l'autopromotion. Nouvelle edition pour les animateurs villageois*, Karthala/GRAAP, 1985

GRET, Lettre du Reseau Recherche-Développement, Dossier-Gestion du terroirs, Paris, no. 10, March 1989

Gubbels, P., 'Peasant Farmer Agricultural Self-development', *Participatory Technology Development*, IIEIA Newsletter, vol. 4, no. 3, 1988, pp. 11–14

Gupta, A., 'Organising the Poor Client Responsive Research System: can the tail wag the dog?', Paper for the Conference on Farmers and Agricultural Research: Complementary Methods, Institute of Development Studies, University of Sussex, 26–31 July 1987

Harrison, P., *The Greening of Africa*, London, Paladin, 1987, pp. 166–9

Jintrawet, A., Smutkupt, S., Wongsamun, C., Katawetin, R., and Kerdsuk, V., 'Extension Activities for Peanuts after Rice in Ban Sum Jan, Northeast Thailand; a case study in farmer-to-farmer extension methodology', Khon Kaen: Farming Systems Research Project, Khon Kaen University, Thailand, June 1985

Johnson, A. W., 'Individuality and Experimentation in Traditional Agriculture', Human Ecology, vol. 1, no. 2, 1972, pp. 149–59

Kabutha, C. and Ford, R., 'Using RRA to Formulate a Village Resources Management Plan, Mbusanyi, Kenya', *RRA Notes 2*, International Institute for Environment and Development, London, 1988, pp. 4–11

Khon Kaen University, Proceedings of the 1985 International Conference on Rapid Rural Appraisal, Rural Systems Research and Farming Systems Research Project, Khon Kaen, Thailand, 1987

Lightfoot, C., de Guia Jr, O. and Ocado, F., 'A Participatory Method for Systems-problem Research: rehabilitating marginal uplands in the Philippines', *Experimental Agriculture*, 24, part 3, 1988, pp. 301–9

Lovelace, G., Subhadira, S. and Simaraks, S. (eds), *Rapid Rural Appraisal in Northeast Thailand: Case studies*, KKU–Ford Rural Systems Research Project, Khon Kaen University, Thailand, 1988

Maseko, P., Scoones, I. and Wilson, K., 'Farmer-based Research and Extension', *Enhancing Dryland Agriculture*, ILEIA Newsletter, vol. 4, no. 4, 1988, pp. 18–19

Maurya, D. M., Bottrall, A. and Farrington, J., 'Improved Livelihoods, Genetic Diversity and Farmer Participation; a strategy for rice breeding in rainfed areas of India', *Experimental Agriculture*, 24, part 3, 1988, pp. 321–31

McCracken, J., *Participatory Rapid Rural Appraisal in Gujarat: a trial model for the Aga Khan Rural Support Programme (India)*, International Institute for Environment and Development, London, 1988

Norman, D., Baker D., Heinrich G. and Worman F., 'Technology Development and Farmer Groups: experience from Botswana', *Experimental Agriculture*, 24, part 3, 1988, pp. 333–42

Repulda, R., Quero Jr, F. V., Ayaso III, R. B., de Guia Jr, O. and Lightfoot, C., 'Doing Research with Resource-poor Farmers: FSDP-EV Perspectives and Programs', Paper for the Conference on Farmers and Agriculture Research: Complementary Methods, Institute of

Development Studies, University of Sussex, 26–31 July 1987

Rhoades, R., 'Farmers and Experimentation', *Discussion Paper 21*, Agricultural Administration (Research and Extension) Network, Overseas Development Institute, London, December 1987

—— and Booth, R. H., 'Farmer-back-to-farmer: a model for generating acceptable agricultural technology', *Agricultural Administration*, vol. 11, 1982, pp. 127–37

Richards, P., *Indigenous Agricultural Revolution: Ecology and food production in West Africa*, London, 1985

Rochette, R. M. (ed.) *Le Sahel en lutte contre la désertification*, GTZ/Margraf, 1989, pp. 221–38

RRA Notes no. 6, The Sustainable Agriculture Programme, International Institute for Environment and Development, London

Scheuermeier, U., 'Approach Development: a contribution to participatory development of techniques based on a practical experience in Tinau Watershed Project, Nepal', Lindau, Switzerland

Scoones, T. and Pretty, J., 'Rapid Rural Appraisal for Economics: exploring incentives to tree management in Sudan', International Institute for Environment and Development, London

Sumberg, J. and Okali, C., 'Farmers, On-farm Research and the Development of New Technology', *Experimental Agriculture*, 24, part 3, 1988, pp. 311–20

Tung, L. and Balina, F. T., Proceedings: Philippines Uplands Research and Extension Training Workshop, ATI, Philippines, 19–24 June 1988

Valmayot, R. V. and Mamon, C. R., 'Research Information Systems for Agriculture and Natural Resources in the Philippines', International Service for National Agricultural Research Management, International Workshop on Agricultural Research Management, The Hague, October 1987

Wolf, E. C., 'Beyond the Green Revolution; new approaches for Third World Agriculture', *Worldwatch Paper 73*, Worldwatch Institute, Washington DC, 1986

4

Farm System and Household Economy as Frameworks for Prioritising and Appraising Technical Research: A Critical Appraisal of Current Approaches

Diana Hunt

Introduction

The focus of technological research and the evaluation of its output are both strongly influenced by the intellectual framework (or paradigm, in the original Kuhnian sense) of the researcher (Chambers and Toulmin, herein). While in Chapter 3 Chambers and Toulmin argue the case for a new research and extension paradigm for complex, diverse and risk-prone (CDR) agriculture, exploring in particular detail its methodological implications, the present paper explores a related theme. This is the need, particularly in CDR agriculture, for an intellectual framework that can facilitate analysis and interpretation of, and hence appropriate technological research planning and evaluation for small-scale multi-enterprise production systems based largely on family labour. These multi-enterprise production units are far from purely agricultural. Indeed, in sub-Saharan Africa farm household members are typically part-time farmers, and are simultaneously engaged in a variety of non-farm productive activities

Some of the ideas discussed in this chapter were first presented at a RUPAG seminar at IDS, Sussex in the autumn of 1985. I am grateful to Mick Howes and Martin Greeley for their comments at that time. I am also grateful to John Dixon of FAO, Rome, for more recent discussion of some of the issues raised, and for the comments of fellow participants at the 1989 Technology and Rural Change workshop (see Preface). Responsibility for the content is, however, solely mine.

both for their own consumption and for the market. Yet the analytical frameworks that have been developed to explore the performance of these units reflect, on the production side, a strong bias towards the agricultural branches of household productive activity. Can they be adapted to give equal weight to non-farm branches of production? The following discussion seeks to answer this question, with particular reference to the need for an appropriate framework for the identification of technological research needs and the appraisal of technological research results.

The discussion begins with a critical review of the approaches and frameworks that have most influenced the analysis of farm household economy in the 1970s and 1980s. This is followed by a proposal for an alternative perspective on the household economy. The chapter concludes with a review of the macro-economic arguments for adopting such a perspective and the identification of the main constraint to following through its implications.

To many it may seem self-evident that rural households in Africa base their economic welfare on a range of productive activities: in self-employment and wage employment, producing for subsistence and the market, both within and outside agriculture. This, of course, is not unique to Africa, but applies in rural areas in both the developing and more developed world (on the latter, see Shuck-smith et al., 1989). For the majority, particularly in the developing world, this diversity is a function of economic necessity; for a more fortunate minority, chiefly in wealthier economies, a much greater element of free choice is involved.

Yet despite this diversity of productive activity, rural households in sub-Saharan Africa have typically been viewed by policy-makers in both the colonial and past-colonial eras as purely agricultural units. This *a priori* assumption has serious implications for the appropriateness of farm development policy. For example, attempts to raise family farm output may fail because the opportunity cost of withdrawing labour from non-farm uses may not be recognised (e.g. Low, 1986a), and households have other, unrecognised, resource-use priorities. Thus a scheme to expand fodder grass production in Baringo District, Kenya, ran into trouble because women wanted to use the grass as thatch (O'Keefe, personal communication, 1989). On the other hand, opportunities to raise farm output may be missed because the scope for expanded non-farm production to raise cash for investment in farming may not be recognised.

For purposes of policy design, abandonment of the *a priori* assumption that rural households are purely agricultural units raises two questions: Can these multi-enterprise households justifiably be viewed as integrated economic units; and if so, how is it appropriate to approach the analysis of their economic performance?

Should sub-Saharan Rural Households be Viewed as Integrated Economic Units?

Three factors are usually held both to constitute the key factors that bind a household together and to represent the primary defining characteristics of households in general:

1 common residence,
2 joint organisation of productive activity,
3 consumption from a common pool of resources.

Traditionally, economists have not questioned these defining characteristics and have represented the household as if it were a totally integrated and harmonious decision unit. Yet as many have pointed out in recent decades – particularly those with an interest in gender issues – this is at best an incomplete and at worst a misleading representation, and one that often fails to specify key factors influencing resource allocation by individual household members (e.g. Guyer, 1986; Evans, 1989). In a typical household individuals generally have some (varying) autonomy in resource allocation. Both this autonomy, which usually carries with it specific responsibilities, and the intra- and extra-household constraints upon it, are important determinants of the scope for changes in resource allocation in particular enterprises and activities undertaken within the household.

Moreover, residential units do not *necessarily* contain a significant element of commonality in resource use at all. In the 1940s, Fortes (Fortes, Steel and Ady, 1949) reported absence of common boundaries to residence, consumption and production units among the Ashanti (Guyer, 1986: 93). For Guyer an important implication of the former discrepancies is the need to recognise that gender may be a more significant factor in the determination of control over resources than the sometimes artificial concept of the household economy.

However, in most sub-Saharan rural communities it would appear that there is a significant element of commonality in household resource allocation, even while individual members retain varying degrees of autonomy over resource use. This commonality, governed as it is by conventions and traditions associated with marriage and kinship ties, is part of what serves to bind the household together. However, these conventions and traditions are not immutable, and the degrees of both joint participation and individual autonomy may be open to renegotiation, for example in the face of new, exogenously introduced, production opportunities.

In sum, five propositions concerning the general nature of the rural household economy in sub-Saharan Africa underlie the analysis that follows:

1   The household is jointly a unit of production and consumption.
2   In the pursuit of joint and individual welfare, tradition, consensus, affection, conflict and compromise all influence the allocation of household resources.
3   In stable situations, in which production opportunities and cultural norms are constant, tradition, consensus and compromise are likely to sustain economic harmony within most households. Tensions are most likely to arise when the production environment is itself changing.
4   Household members pursue their individual welfare subject to overall household resource constraints and existing norms concerning intra-household resource allocation.
5   Most household members have a sense of joint welfare and a shared interest in the increased welfare of one or more members, provided that this does not entail an involuntary absolute loss in material welfare for the others.

Past and Present Approaches to the Theorisation and Modelling of Farm System and Household Economy

In this section we review those approaches and frameworks that in recent years have most influenced the analysis of resource use by African farm households. Given our primary interest in technical change, we start with farming systems analysis.

The Farming Systems Approach to the Analysis of Farm Households: Background, Distinctive Features and Notable Omissions

As Chambers and Toulmin note (Chapter 3), farming systems research (FSR) emerged in a context in which the dominant approach to agricultural research and extension in the Third World had previously been mono-crop- and often mono-practice-

52

oriented.[1] There had been two dominant reasons for this: the colonial and post-colonial preoccupation with the expansion of production of key export and food crops, and the need for a simple extension message for dissemination by personnel with limited formal education. Meanwhile, the promotion of rural development took place in a context dominated by two presumptions: that the chief basis for the expansion of market-oriented rural production was agriculture, and that the typical pattern of agricultural evolution is towards increasing regional and farm-level specialisation. Agricultural research and extension had generally ignored the broader ramifications, both for the rest of the farm and for household consumption, of modifying a particular cropping practice and/or the enterprise mix. Likewise the resource constraints and the particular goals (not necessarily yield or profit-maximisation) of small farmers were generally ignored. It was the recognition of the need to take account of these factors that led to the emergence of FSR.[2]

FSR is not in itself a form or method of innovative technological research. Rather it constitutes an approach to the (multidisciplinary) analysis of existing farm systems, intended to provide an improved framework for the identification of potential innovations and of research priorities, and the evaluation of proposed innovations. Among the distinctive features claimed for FSR are:

1   a concern with helping the small farmer;
2   an attempt to understand the farmer's existing resource allocation behaviour;
3   recognition of the soundness of many traditional farming practices;
4   emphasis on farmer involvement in identifying research needs and testing results;
5   collaboration of technical and social scientists, together with farmers, in the effort to raise output;
6   recognition of the need for a holistic approach to the analysis of the complete farm system.[3]

1. When the new ('Green Revolution') grain hybrids were developed, it was deemed necessary to provide farmers with a package of recommendations for the most yield-sensitive operations. However, the basic approach remained the same: single crop and yield maximisation-oriented, with a sequence of research trials for individual operations until a set of recommended practices was generated.

2. See, e.g., the pioneering article by M. Collinson, 'The Evaluation of Innovations for Peasant Farming', *East African Journal*, 1, 2, July, 1968; also D. Norman, 'Initiating Change in Traditional Agriculture', *Proceedings of the Agricultural Society of Nigeria*, 1970.

3. The underlying philosophy of farming systems analysis is further elaborated in

Various authors have pointed out that farming systems analysis has much in common, at both the philosophical and methodological levels, with farm management research conducted in the United States in the early decades of the twentieth century.[4] However, within developing countries the approach that evolved in the 1970s was an innovatory and important attempt to make research relevant to resource-poor farmers. A methodology was worked out, which was designed to identify appropriate technical priorities for farm trial or on-station research. This was done through a sequence of selective interviews with key personnel, and informal and formal farm survey, and was conducted by a multidisciplinary team of physical and social scientists.

With growing experience of implementation, FSR has recently come under attack for a variety of reasons, including:

1 Farmer insensitivity: in practice, farming systems researchers have continued to assume that the flow of technical information and advice should be basically one-way (see Chambers and Toulmin, herein Chapter 3).
2 Failure to consider the gender implications of technical change (e.g. Poats et al., 1988);
3 The development of cumbersome, costly, multidisciplinary research teams (Chambers and Toulmin, Chapter 3).
4 Generation of results, which are regionally specific and not widely generalisable.[5]
5 A focus largely restricted to crop production, with little attempt to include livestock enterprises.[6]

Chambers and Toulmin (Chapter 3) have made an in-depth analysis of the implications of criticisms 1, 3 and 4. The latter has been used by some to suggest that FSR can never be cost-effective. However, this depends partly on whether criticism 3 is met, while it is also necessary to allow sufficient lead-time for appropriate new research to develop. As the stock of basic knowledge and of potentially viable innovations accumulates, so the cost of and

---

D. Norman et al. *Farming Systems in the Nigerian Savannah*, Boulder, 1982, pp. 4, 5; also N. Shaner et al., *Farming Systems Research and Development*, Boulder, 1982.

4. See e.g. Eicher in D. Norman, 'The Farming Systems Approach', Michigan State University, 1980; E. Gilbert et al., 'Farming Systems Research', Michigan State University, 1980.

5. See Norman, 'The Farming Systems Approach'; also Gilbert et al., 'Farming Systems Research', pp. 104, 107, 113.

6. See Norman, 'The Farming Systems Approach'.

lead-time for generation of effective innovations is likely to decline.[7] However, even if these criticisms are met – even if methodologies for farmer – farmer and farmer – researcher interaction are improved, crop and livestock research increasingly integrated, the stock of potentially viable innovations increased, as also communication between research units in different countries but similar ecological zones – there will still be an outstanding problem, which neither improved farming systems research nor the more emphatically farmer-sensitive 'farmer-first-and-last' approach can alone resolve. This problem derives from four related facts:

1 the accepted objectives of stabilising and raising the incomes of poor rural households;
2 the fact that for many of these households in the Third World only about half of total income comes from farming;
3 the widely recognised fact that the farm system competes for a common pool of resources (both labour and capital) with other branches of the household economy;
4 the less widely acknowledged fact that there are positive as well as competitive links between different branches of the household economy.

Evidence for the second of these points is available from a number of sources. Chuta and Liedholm (1979) report statistics for the share of non-farm income for rural households in five Third World countries (two African and three Asian) at between 22 and 43 per cent.[8] In Kenya (not included in Chuta and Liedholm's figures) estimates range from 35 to 50 per cent.[9] Matlon (cited also in Lipton, 1989) reports figures between 20 and 40 per cent for three villages in northern Nigeria. So far as I am aware, however, all of these figures are underestimates, because they exclude any attempt to value the welfare generated by labour-time devoted to household tasks, whose use-value to the household may none the less be very high.[10]

7. It is worth remembering that in the United States it took twenty-three years of research before the first hybrid maize seed was ready for distribution, and at least ten years for hybrid sorghum (see Z. Griliches, 'Research Costs and Social Returns', *Journal of Political Economy*, 1958, pp. 424 (Table 1); 428.

8. E. Chuta and C. Liedholm 'Rural Non-farm Employment', Michigan State University, 1979, p. 7. The African countries cited were Nigeria and Sierra Leone, with Pakistan, Taiwan and South Korea from Asia.

9. See I. Livingstone *Rural Development, Employment and Incomes in Kenya*, Addis Ababa, 1981, p. 14.20; D. Hunt, *The Impending Crisis in Kenya*, Aldershot, 1984 p. 39; J. Carlsen, *Economic and Social Transformation in Rural Kenya*, Uppsala, 1980 pp. 174, 175.

10. Indirect evidence for this last point comes from the substantial proportion of

Competition for resources occurs between productive activities, between current production and consumption, and between these and investment in the expansion of future output. Complementarities between different branches of the household economy also arise in various forms. For example, an increase in labour productivity in one branch – food processing, say – may result in the release of increased labour to another; alternatively, cash earned in one branch may finance investment in another.

That the farm system forms but one part of the larger household economy is acknowledged by farm systems analysts,[11] as well as by advocates of the farmer-first-and-last approach,[12] but the implications have only been partially followed through. Generally, these approaches take three features of the full household economy into account in recommendations for research design: (1) production goals, as these affect farming, (2) current levels of resource commitment to the farm system, and (3) risk-aversion. In application though, both approaches have remained focused on the farm branch of the household economy.

While it is true that FSR has contributed, through its emphasis on labour-time budgets, to the recognition of the non-farm labour demands of the farm household – particularly those for household maintenance and reproduction – and that farming systems-oriented technical research in various countries has been extended to include items such as crop storage and on-farm production of woodfuel, such developments have been limited to certain non-farm activities only, and may be seen as logical extensions of traditional concerns with crop production. The woodfuel problem has also given rise to the development of separate research institutions and programmes concentrating on issues that are still underemphasised or ignored in agricultural institutions and programmes.[13] However, there are other branches of rural household productive activity (both for own-use

---

household labour-time, female in particular, devoted to these tasks, combined with the presumption that family labour is rationally allocated on the basis of subjective estimates of marginal utility. In fact, it is obvious that activities such as fetching water and firewood, and food processing and preparation are essential to household maintenance.

11. See e.g. Norman, 'The Farming Systems Approach', p. 2; Gilbert et al., 'Farming Systems Research', pp. 2, 6; FAO, Farm Management and Production Economic Services, *Farm Systems Development*, Rome, 1989, p. 14.

12. See, e.g., Chambers, in C. Conroy and M. Lipsinoff, *The Greening of Aid*, London, 1989.

13. For instance, the Kenyan Woodfuel Development Programme and the various cooking stove projects that have evolved in different parts of sub-Saharan Africa.

and for the market), which have looser linkages with agriculture and which have consequently received even less systematic attention, for example, production of roofing materials (though this has been the subject of privately sponsored R & D),[14] food processing, charcoal production, handicrafts, etc.

What is needed is an analytical perspective that gives equal weight to all branches of the household economy in an attempt to establish where the priorities for technical research should lie. As the debate on gender issues has shown, this should also take into account intra-household resource allocation between persons. In the next section we consider the extent to which available models of the household economy, particularly those developed to represent rural households in the Third World, can perform these functions.

Models of the Household Economy: Becker, Chayanov, the World Bank/Stanford Approach and Low

In terms of conception and analytical focus, household economic modelling is of a different order from farming systems analysis. For a start, such models have not been designed with the specific objectives of providing a framework or basis for the identification of technical research priorities and research appraisal. However, these models have been the focus of much attention, and in their various forms have been used to explore a number of issues relating to technological change. As we shall see, they provide some scope for fulfilling the roles just noted, and also reveal varying degrees of complementarity with farming systems research. We start by looking briefly at two earlier approaches to household economic theorisation, which have influenced more recent analysis of rural household economic behaviour.

The 'New Household Economics' We referred earlier (p. 56) to the emphasis given to time budgets in farming systems analysis. Becker (1965) places even stronger emphasis on labour-time as *the* household production constraint in his theory of the allocation of time. This theory, like all household economic theory, is based on a perception of the household as a joint production and consumption unit.

On the production side, Becker characterises households as combining time and purchased goods to produce those goods that directly enter their utility function. These goods (denoted as Z

14. See the advertisement in *Appropriate Technology* 16(1), June 1989; and the assessment by S. Mwangi, 'Shelter Sure', *Rural Development in Practice*, 1, 4, May 1989.

goods) are the ultimate source of well-being; they include such things as leisure, nourishment, entertainment, sleep, physical warmth and children. These are produced through varying combinations of money expenditure and direct time input. There is, however, only one basic constraint on raising household well-being: time. More money can be generated by committing more time to earning it. Thus time is the basic productive resource, and the basic resource constraint.

For Becker all time at the disposal of an individual has a money value, whether or not it is used to generate money income. If it is not, but instead is used for non-market production or consumed as leisure, then its value is its 'opportunity cost': the money income forgone by not engaging in money income-generation. From this stems a number of corollaries: for example, since the value of time (in work and leisure) is directly related to the money income generating capacity of the individual involved, the value of time varies between individuals, both between and within households. Secondly, since all time can be valued in terms of opportunity cost, all Z goods, including leisure, can be valued as the sum of the values of the purchased goods, time and other own-produced goods (in turn, the product of time and purchased inputs) used in their production.[15]

A primary focus of the new household economics is on household technology choice for the production of Z goods. This choice is manifested in two sets of decisions: the extent to which different household members allocate their time directly to Z goods production at home or to earning money; and the types of capital goods and recurrent inputs which the household purchases for Z goods production. Thus, a household in a developed economy may produce clean clothes by sending them to the laundry, taking them to the launderette, using its own washing machine or by hand-washing. Each method requires different combinations of time, recurrent cash outlay and use of home-owned capital goods. The choice will depend on the time and other costs involved in each method, and on the value of this time and associated costs.[16] As the value of time rises, so a more capital-intensive, labour-saving

15. One point, which Becker does not draw out here, is that it follows that the values of all home-produced goods vary between and within households according to the money income earning capacity of the producer.

16. A feature of this perspective is that it assumes that an individual could obtain the same marginal income whatever the number of days and hours worked. In practice, this is not necessarily the case.

method is likely to be chosen. Becker assumes, in the neoclassical tradition, that a range of technical options for the production of Z goods exists.

The 'new household economics' was developed largely with the urban households of North America in mind. However, the approach has also influenced farm household modelling in the Third World. Before turning to consider how, we turn briefly to the pioneer of farm household economic theory, Chayanov.

Chayanov's Model of Peasant Household Resource Allocation   Chayanov, writing in the 1920s,[17] was among the first to recognise the joint roles of the peasant household as a production and a consumption unit, and to recognise too that household production may be geared to own consumption and to the market. There are, however, significant differences of approach between Chayanov and Becker, some of which are complementary, some not. In the former category is Chayanov's emphasis on the role of the demographic composition of the household, both as a determinant of ability to meet basic consumption needs (largely due to variations in the producer:consumer ratio) and as a source of rising labour productivity, due to increasing specialisation as the absolute number of producers rises.

However, Chayanov differs from Becker in his specification of the household decision calculus. For Chayanov, this is based on a subjective evaluation of the marginal utility of output compared with the marginal disutility of the effort entailed in its production. Maximising the *value* of household 'full income'[18] is not possible, primarily because labour used on the family farm is unwaged.[19] Chayanov does not consider the possibility that household labour might be valued by attributing to it the market wage, presumably because in his view most peasants only put their labour on the market in periods of economic crisis, such as harvest failure or market collapse, when the wage is abnormally low. Chayanov also emphasises that household subsistence production is not price-responsive.[20]

17. See D. Thorner et al. (eds), *A. V. Chayanov on the Theory of Peasant Economy*, Illinois, 1966.

18. A concept introduced by Becker; it measures the value of all the Z goods, including leisure, available to a household over a given period.

19. In so far as the household produces commodities for the market, the net return obtained is a joint return to capital and labour, i.e. the return comprises joint wage and profit elements which cannot be distinguished.

20. That is, relative prices and price changes do not affect the household's estimate of the utility to it of these items.

The Chayanovian perspective, with its allowance for greater detachment of the household economy from the influence of market prices, by implication allows a greater role for custom and convention in the determination of household production patterns. In contrast, in the 'new household economics' intra-household allocation of productive activity is explained solely by the opportunity cost of household members' time. Implicitly, custom and convention are assumed to reinforce this, not to conflict with it.

On the other hand, certain influences on the operation of the household economy, including (1) the role of risk (natural and market); (2) the intrahousehold distribution of control over productive resources; and (3) the role of custom and convention in resource use, are explored by neither Becker nor Chayanov.

Recent Developments in Household Economics with Reference to Farm Households in Less Developed Countries   Among the more recent developments in the theorisation of household economic behaviour, which have particular reference to rural households in the Third World, two main lines of approach can be identified. One is that adopted by the World Bank and collaborating institutions (and/or following the same tradition); the other is that developed in Low (1986a), with specific reference to sub-Saharan Africa.

The approach to household equilibrium modelling adopted over the last two decades, most notably at the World Bank and the Food Research Institute, Stanford University, falls in the mainstream of neoclassical economic theorising. Singh, Squire and Strauss (1986a) summarise two versions of the approach, of which one – version A – is preferred, due to greater ease of modelling. Both versions emphasise the importance of capturing in a single model the joint role of the household as a production and consumption unit in order to inform policy formation appropriately. The household is assumed to be a rational utility-maximiser.

According to version A, the household is a price-taker in all markets, for all commodities, including leisure. Labour too can be valued at the prevailing market wage. In these circumstances, a utility-maximising household should logically proceed by first maximising its income, given prices, its resource endowment and production technology, and then proceed to its consumption and leisure choices. There is still an important interaction between production and consumption decisions, for a change in the former will affect the latter via the impact on total household income.

In version B the assumption that all prices are given to the household is dropped, and it is acknowledged that some household production decisions may be made independently of market prices. For instance, a household may choose to equate domestic production and consumption for some commodities, perhaps because of the non-existence of a market, or because sales and purchase prices differ for an identical commodity (Singh et al., 1986a: 155). In this case production and consumption decisions are simultaneously determined (which makes mathematical modelling more difficult).

In both versions Singh et al. specify a cash cost-constraint on consumption, but no such constraint is specified for production. Instead, production is constrained by a fixed land endowment. Yet three factors may generate such a constraint: lack of access to, and/or the cost of, credit; the competing claims of consumption demands, which must be met in order to maintain the farm labour force; and the time-lag involved in agricultural production, which means profit-maximising resource allocation today generates income weeks, months or even years into the future. In the intervening period the household must meet its basic maintenance needs, and these are likely to require some minimal cash outlay. The value of current income may, therefore, exercise a constraint on production. This omission does not necessarily invalidate a key empirical conclusion emphasised by Singh et al. (1986a: 163), which is that marketed output always responds positively to own price, since it may be possible to expand output without increased expenditure in the first instance by reducing the consumption of leisure. However, this is least likely for the poorest households.

The types of model described by Singh et al. have been used primarily to estimate production price-response and consumption demand elasticities.[21] On the production side, the focus has been on farm output, and has virtually ignored non-farm production (only the 'production' of leisure is included), although in principle it should be possible to extend the analysis to other market-oriented enterprises.[22] Choice of production technique has only arisen as a significant issue in so far as it is price-responsive (e.g. price elasticity of demand for fertiliser). The models have not been used for the identification of technological research priorities. The emphasis has been on estimating

21. The modelling techniques used have been of two types: programming and econometric estimation.

22. Certain extensions of the use of such models have, however, also been explored, for example, with respect to the impact on nutrition and health of changes in household production and consumption: see J. Strauss, 'Joint Determination of Food Consumption and Production in Rural Sierra Leone', *Journal of Development Economics*, 14, 1984; and Pitt and Rosezweig, in I. Singh et al., *Agricultural Household Models*, Baltimore, 1986b.

existing production functions, rather than transforming them.

Low's Model of the Farm Household Economy   Low (1986a) presents an alternative farm household model, developed in reaction to the types just reviewed, which attempts to reflect more accurately the structural conditions of African farming systems. He rejects three of the assumptions upon which standard, neoclassical farm household modelling is based, viz.:

1   Land area is fixed for individual farms (whereas in much of sub-Saharan Africa it still is not).
2   There are diminishing, as opposed to constant, returns to labour on the family farm (less likely when land is variable).
3   Households face a constant and uniform market wage for all members (in fact, wage opportunities vary by age and gender).

Low also considers the potentially positive per capita output impacts of a rising number of productive members per household, due to scale economies and exploitation of opportunities for specialisation. His alternative model is inspired jointly by Becker and Chayanov. From Becker, Low adopts the propositions that:

1   Farm households are producers of Z goods.
2   Households face various technical options for Z goods production.
3   Households usually require the member(s) with access to the lowest market wage rate to undertake domestic production of Z goods, unless this lower opportunity cost is more than offset by their lower productivity.

Because household labour productivity in Z goods production may be positively related to the number of productive members in the household, households of different size may face different trade-off rates between own production and purchase of certain goods.

Low also reminds us that conditions of communal land tenure in much of sub-Saharan Africa tend to keep down the cost of own production of certain types of goods such as water, grazing for livestock, thatching grass and woodfuel. This is because under traditional tenure an individual with cultivation rights also has free rights of access to uncultivated land to obtain these resources.

The main focus of Low's application of his model is, however, once again in the area of farm production, with a particular focus on farm resource allocation and productivity. Thus, restriction of hybrid maize adoption to subsistence production on many farms in Swaziland is explained by the fact that it raises the net productivity of those household members with relatively low labour-time opportunity costs who are already engaged in subsistence production – thereby releasing some of their time to other uses – but does not raise above its existing level the net productivity of those whose labour-time is commercially-oriented.

Low's model synchronises with the basic farming systems approach (see pp. 53, 54 above). As Low (1986b) emphasises, it can strengthen farming systems analysis in several ways: by providing an analytical framework for the analysis of labour allocation and the household's labour-time constraint, and by pinpointing the significance for innovation adoption of the different values of households members' labour time; linked to this are variations in household size and composition. Indeed, Low (1986b) argues that in defining technical recommendation domains, farms should not just be grouped by physical resource endowments but also by key economic characteristics, most notably labour force size and access to cash income. These contributions appear to be relevant also to the farmer-first-and-last approach.

Recent Farm Household Modelling: A Reappraisal

Both recent mainstream farm household models and Low's modified version represent significant advances in the attempt to come to grips with the resource allocation behaviour of joint production and consumption units based on the domestic group: the former most notably in their exploration of the interlinked production and consumption changes that are induced by price variations; the latter through the development of a productive resource allocation model that more accurately reflects the structural conditions of the African farm household. Neither, however, provides a fully satisfactory framework for the analysis of rural household economic systems in sub-Saharan Africa. In particular, we may note the following omissions from both perspectives:

1　failure to focus, on the production side, on the full household economy;
2　failure to explore potential dynamic interconnections between the different production branches of the household economy;
3　failure to consider the possibility that the intra-household division of labour may sometimes be determined not by relative productivity, but by conflicting convention (Evans, in Poats et al., 1988);
4　failure to explore possible rigidities in intra-household resource

use, due to the relative autonomy of individual members in controlling particular resources;

5   failure to consider intra-household tensions, and possible inequities, arising both in resource allocation for production and in the distribution of consumption;

6   treatment of each individual's household labour-time as homogeneous in effort and drudgery expended, and as having a given marginal product in any line of production, irrespective of the timing of the labour input. The latter does not apply in farming where the marginal product of labour is highest at the busiest periods of the cropping season. The former both vary according to the task, to the length of time worked, to human choice concerning degree of effort expended and to energy availability, which may decline in the hungry season (Levi, 1987).

In addition, both approaches have given little emphasis to the role of risk (natural and market) in influencing household production decisions. And in their analytical approach, both rely on the attribution of monetary values to all family labour-time.

It is with the first two of these criticisms that the rest of this chapter is primarily concerned. Criticisms 3–5 have already been discussed in a number of sources (see e.g. Whitehead, in Young et al., 1981; Poats et al., 1988; Evans, 1989). It is noteworthy too that criticism 6 raises the question of whether an individual's labour-time can be valued reliably in terms of a notional 'normal' hourly rate of farm income (see also note 16). Chayanov's emphasis on the significance of the subjective valuation of household labour-time does not give rise to these problems. The implication of emphasising such subjective valuation, and/or of allowing that households may recognise real opportunity costs in terms of output forgone when reallocating resources, but with no simple, objective basis for valuing these, is that it is necessary to rely on households' own observations as to where the primary constraints to raising their welfare lie.

The Household Economy in Sub-Saharan Africa

In this section I suggest an alternative analytical perspective, which might serve as a starting point for the analysis of rural household economies. This will help to provide a framework for, *inter alia*, the identification of technological research priorities and the evaluation of technical innovations.

Bearing in mind the need for sensitivity to intra-household constraints on resource access and use, I suggest that analyses of rural household economy could usefully start from the five propositions concerning the nature of household economy outlined above (p. 52) combined with the following:

1 Household members divide their time between various productive activities and leisure. The former may be categorised as:
   (a) farming:
       for own consumption,
       for market exchange.
   (b) off-farm commercial activity:
       wage employment,
       production of goods and services for sale.
   (c) household tasks (including production of handicrafts for own use).
2 Low-income farm households maximise utility with partial disregard to market prices. Infinite utility is attached to meeting basic consumption requirements, and preference is given to doing so in a risk-minimising manner.
3 Choice of production *method* is, however, usually cost-sensitive, taking into account risk, whether production is for basic consumption or is over and above this (although costs are not necessarily given money values).
4 Whether or not there are other fixed resource constraints (e.g. land, irrigation water) the basic resource to which the household seeks to maximise returns is labour-time. (These returns are either subjectively or objectively estimated according to whether production is price-sensitive.)
5 Returns to time may be maximised partly through the use of time to generate multiple products (or utilities). For example, time spent fetching water may also be a social activity; while participation in certain social activities may also bring potential future economic benefits (cf. Jones, 1968; also Hunt, 1977: 86).
6 The productive resources used in these activities are often highly fungible.[23] Hence, simple expansion of production in one activity, if it implies use of an already fully utilised resource, will imply reduction of production in another with, possibly,

---

23. Recognition of this fact, together with the utility-maximising and multi-enterprise features of the household economy, is sufficient to provide a rational explanation of the failure of some small farm credit schemes, with high rates of resource diversion to non-farm use.

consequent changes also in the pattern of household consumption. Conversely, productivity increase in one branch may release resources for output expansion in another also. (Such productivity increase may or may not entail some initial resource competition.)

7 Linkages between production and consumption may also be both competitive and complementary.

The multi-enterprise rural household engaged in several lines of commercial activity as well as in production for its own needs is becoming increasingly recognised as a significant economic phenomenon, not only in the developing world but in parts of the developed world also (Shucksmith et al., 1989). However, in the case of the developing world, such recognition has generally been geared to informing *agricultural* policy – and in particular agricultural research – with a view to making this more sensitive to household objectives and to the resources that households can make available to farming. The thesis of this chapter is that a less biased perspective might generate other research priorities, offering the prospect of higher welfare gains and, more specifically, the possibility of increased farm as well as non-farm output. The following diagrammatic analysis of the circular flow of resources within the household economy seeks to illustrate this point.

Figures 4.1–4 assume that the household is economically harmonious, with no serious unresolved conflicts over resource allocation, having what is internally accepted as an equitable distribution of consumption. In each figure household resource availabilities are given at the top, the various forms of resource use are represented on the left-hand side of each circular flow chart, the alternative forms of disposition of household income are shown at the bottom, and the impact of increased income upon resource availabilities are recorded on the right-hand side.

Figure 4.1 illustrates the household in a context of no technical change. Figures 4.2–4, in contrast, provide three examples of the way in which technological change in one branch of productive activity may generate a variety of positive repercussions in other branches.

Figure 4.2 illustrates the 'first round' potential benefits arising from improved labour productivity in a household maintenance task – water-fetching – and one way in which this might indirectly impinge on the household's farm production via an improvement in household health and nutrition. Increased productivity in household maintenance might also release additional labour-time to

Figure 4.1   The 'circular flow' of resources in the household economy
1. The stable flow

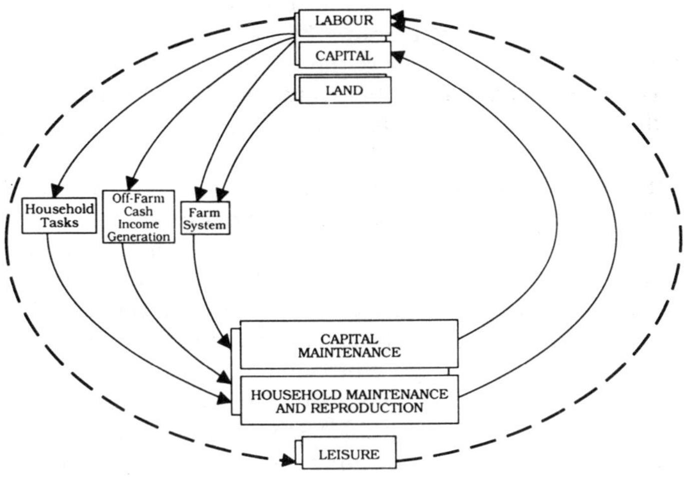

Figure 4.2   The 'circular flow' of resources in the household economy
2. The cumulative spiral, Example 1

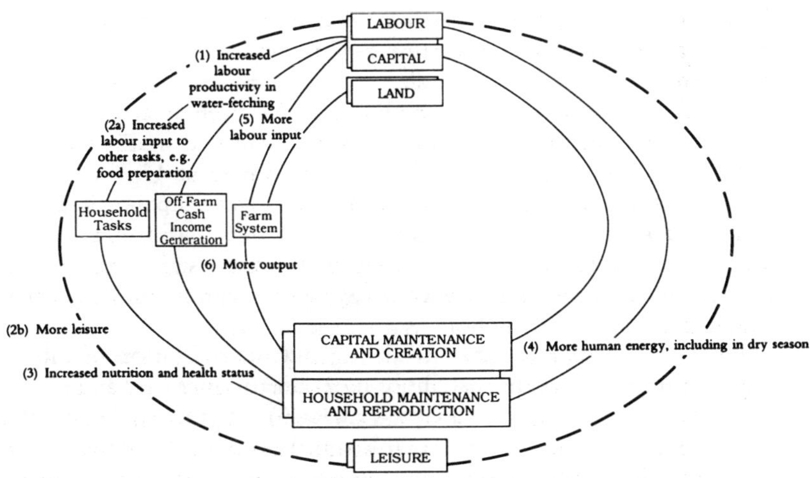

Note: Numerical sequence refers to order of events.

non-farm commercial production, with some of the cash returns feeding through into farming (Figure 4.3). Or an upward output spiral might be initiated through increased labour productivity in commercial non-farm production. This could apply, for example, when apparently slack labour in the hungry season could be used, but due to hunger is not, to overcome a subsequent energy-intensive farm labour bottleneck (e.g. land-clearing), and where development of non-farm cash-income generation permits extra food purchase to release this energy constraint.

Figure 4.4 illustrates one particular instance of the potential impact of increased labour productivity in farming itself. This is similar to the situation described by Low, in which, given a sufficient discrepancy between food sale and food purchase prices, it pays the rural household to produce for own consumption but not for the market (p. 62). The presumption reflected in Figure 4.4 is that following a particular agricultural innovation, there is still no branch of commercial agriculture accessible to the small farmer that brings as high a cash return to labour inputs as off-farm market-oriented activity (including the sale of labour-time). However, the household does produce its own subsistence requirements.

None of the preceding schematic representations allows for the possibility that the household may be characterised by internal tensions regarding access to productive resources and/or the distribution of consumption. Nor do they allow for the possibility that the outcome of such intra-household tensions may be a matter of policy concern. If, in fact, these two sets of conditions prevail, then they too can be expected to influence both the identification of research priorities and innovation evaluation.

In this regard, a similar, albeit more complex, diagrammatic approach can be used to explore linkages between activities classified by the controlling gender. This is illustrated in Figure 4.5, which uses the example, reported for north-west Uganda, that women may brew beer to pay for labour or cash inputs for male-controlled cash-crop production.[24] The question-marks attached to the possible modes of use of the increased income signify that there may be significant variation both between and within societies in the extent to which the household as a whole benefits.

Figure 4.5 does not explicitly bring out the operation of inter-gender conflicts over the distribution of control over land and other assets (although it could be adapted to do so). In practice, however, such tensions may have considerable significance for both the scope

24. J. Harmsworth, personal communication, September 1989.

Figure 4.3   The 'circular flow' of resources in the household economy 3. The cumulative spiral, Example 2

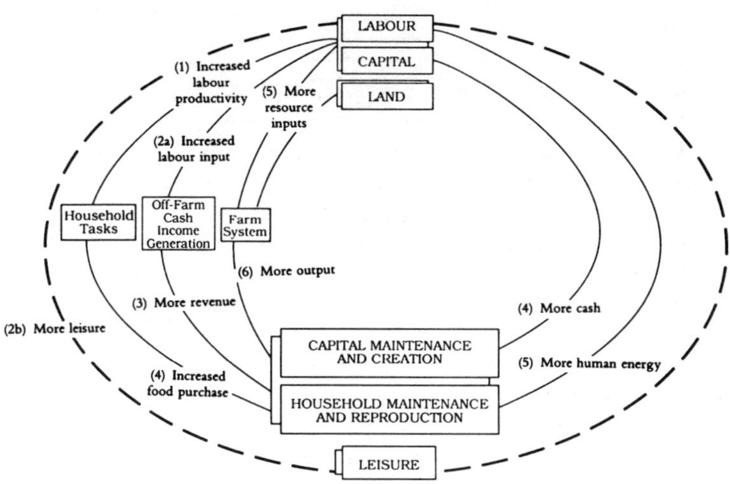

Note: Numerical sequence refers to order of events.

Figure 4.4   The 'circular flow' of resources in the household economy 4. The cumulative spiral, Example 3

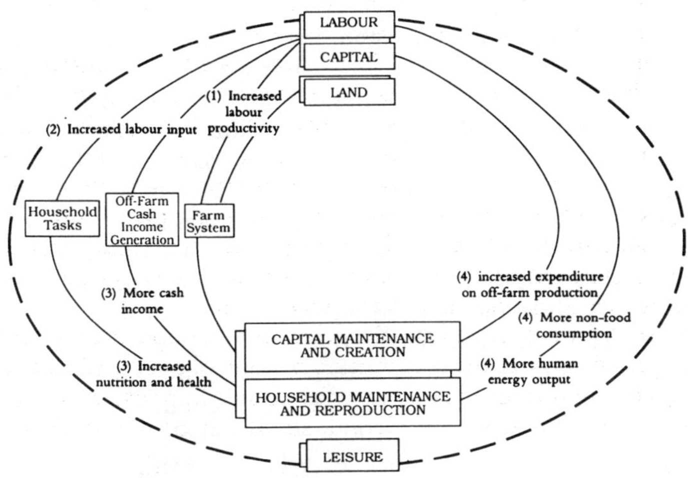

Note: Numerical sequence refers to order of events.

Figure 4.5   Inter-gender production linkages within the household: an example

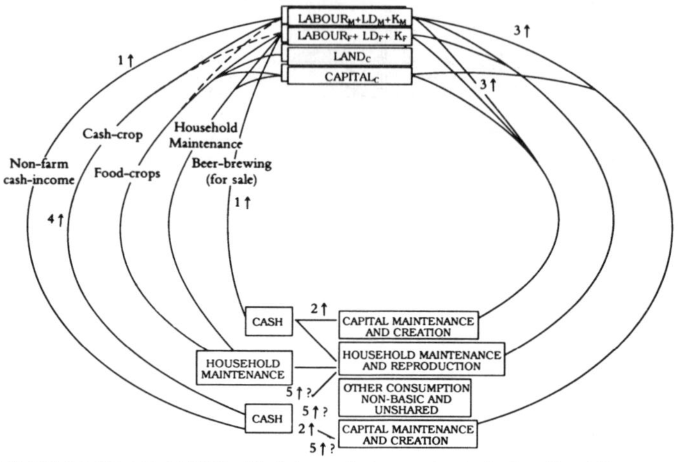

Key: LD=Land; K=Capital; subscripts 'M', 'F' and 'C' refer to ownership of resources by males, females and household common property; ↑ increase in; numerical sequence refers to sequence of events; dashed lines represent input into activites that are primarily the responsibility of the other gender.

for, and impact of, technical change. In parts of western Kenya, for example, women who commit labour to tree-planting to supply woodfuel run the risk that males may assert their traditional rights of tree-ownership in order to appropriate the output (see Bradley et al., 1985).

Figures 4.1–5 illustrate the application of a particular perspective on the household economy to the analysis of resource flows within it. The perspective itself can be applied, both *ex ante* and *ex post*, to explore the likely and actual impact of technical change. It can also be used to explore the impact of other economic changes of significance for the household, such as a change in relative prices. While applied here to low-income rural households, it can also be applied to wealthier and to urban households. It should, of course, be emphasised that the predictive quality of the *ex ante* analysis will only be as good as the data fed into it.

To this end, the basic perspective that we have outlined must first be applied to the collection of relevant empirical information for analysis. It is here, in the area of data collection, as also in that of technical information dissemination to households, that there is a significant potential complementarity both with the 'farmer-first' approach outlined by Chambers and Toulmin in Chapter 3, and with the more time-intensive farming systems approach as originally advocated.

However, the extension of these approaches to data collection on the full household economy would have significant methodological implications: preliminary peer group discussions would have to range over a wider range of issues in order to obtain a clear picture of household objectives and overall production strategy and key constraints; the importance of breaking down some meetings by gender would be increased (although sensitively handled joint meetings can increase inter-gender awareness of gender-specific issues); and time-use studies and economic appraisal of alternative market-oriented activities would have to give unbiased emphasis to farm and non-farm production. Furthermore, such investigations must be sensitive to the way in which intra-household production and production-consumption linkages may determine existing resource use patterns. (Jiggins (1988), citing Huss-Ashmore (1982), gives the example of Lesotho households whose food production and consumption patterns are determined not by preference, risk-aversion, land or climatic constraints, but by fuel availability variations over the calendar year.[25] We also noted (p. 65) that a single time input may generate multiple products. This will clearly affect household members' estimates of the time-cost of producing such products.)

Precisely what methods are used for data collection will be partly a function of the time and other resources available. 'Rapid rural appraisal' techniques, even if they genuinely apply a 'household-first' perspective, cannot initially provide the degree of understanding of rural economy and society that can come from well-conducted, in-depth, time-use surveys. The risk with the latter is that they may be too narrow or unrepresentative in coverage. On the other hand, large-scale, one-off or limited interview random sample surveys – also costly to conduct and analyse – may generate particular problems of enumerator supervision and of remoteness of the data analysts from the surveyed areas. There are few hard-and-fast rules on the *form* of data collection methodology, since each has merits and drawbacks; but there are several on the *approach*. One is that it is essential to have some – as much as possible – knowledge of the area before moving in to conduct any kind of household meetings and interviews. This knowledge can be obtained from published sources and interviews with local experts, as

25. Jiggins, in S. Poats et al., *Gender Issues in Farming Systems*, Boulder, 1988.

emphasised in FSR. The greater such knowledge the greater the likelihood that outsiders 'will know what questions to ask',[26] and that appropriate relative emphases will be placed on single and mixed gender meetings and interviews. The greater also will be the effectiveness with which households' own specification of their needs can be understood and interpreted.

Allowing low-income rural household peer groups a primary role in identifying key production problems, and possible solutions, also requires outside agencies to adopt a more flexible perspective with respect to the appraisal of the benefits of technical change. For the immediate benefits may take the form of time saved or improved diet, rather than quantitatively increased output (let alone *marketed* output). A necessary feature of a 'household-first-and-last' approach lies in recognising both the rights and the efficiency of rural households in deciding how to allocate increased time and human energy in order to raise welfare. At the same time, it is unwise to assume that the initial response will necessarily equal the final adjustment.

Conclusion

Our argument has concentrated on the need to interpret and analyse the household economic unit as nearly as possible 'as it is'. Three macro-level arguments reinforce the case for doing so. First,

> A huge edifice of training, extension effort, credit schemes, often including transport and processing as well as infrastructural investments, may be built around selected technology as the core of a development programme. (Kenya Government, 1977: 1)

If the technology does not fit the intended adopters' priorities and circumstances, it will be unacceptable to them, and not only the research but the accompanying development effort will have been wasted.

Secondly, the multi-enterprise household, combining part-time farming with off-farm occupations (some, but not all, farm-related), has proved, in some of the already industrialised countries and in some of the newly industrialising countries, to be not just the preliminary stage in rural development but rather, an adaptable

26. A. Cheater, observation in Technology and Rural Change workshop discussion, School of African and Asian Studies, Sussex University, September 1989.

and resilient economic life-style. For some it is also psychologically more satisfying, due both to its diversity and its rural base. (Retention of the rural population *in situ* can also serve to contain the social costs of excess rural–urban migration.)

Thirdly, it looks increasingly probable that developing countries, particularly the poorest and those with the lowest agricultural potential (some of which are in sub-Saharan Africa), are going to have to confront a growing international disparity in agricultural technological capabilities (both at the level of research and production) – a disparity arising largely from recent and prospective advances in genetic engineering. This will cause low-productivity agricultural producers to face increasing competition, including in basic food crops, from high-productivity producers with lower unit costs (unless, of course, their governments take protective action). The stronger and more diversified the rural economic base as a whole, the greater the likelihood that rural producers will be able to withstand such pressures.

The main impediment to the implementation of the broader perspective advocated here for identifying technological research priorities meriting public support has been and remains institutional, lying in the structure of existing technological research institutions and in the training programmes and priorities of the higher education centres that produce the local staff for these institutions. Agronomic research is now recognised as a prestige profession in the developing world, engineering research into small-scale, low-cost and labour-intensive 'appropriate' technologies for the non-farm sector usually is not. Here too there is a continuing need for change, including increased recognition of those initiatives already being taken (see Bruinsma and Nout, this volume, Chapter 11).

References

Appropriate Technology 16 (1), June 1989
Becker, G., 'A Theory of the Allocation of Time', *Economic Journal* 75, 1965
Bradley, P., Charangi, N. and Van Gelder, A., 'Development Research and Energy Planning in Kenya', *Ambio* 14, 4–5, 1985
Bruinsma, D. and Nout, M., 'Choice of Technology in Food Processing

for Rural Development', Chapter 11, this volume

Carlsen, J., *Economic and Social Transformation in Rural Kenya*, Scandinavian Institute of Social Studies, Uppsala, 1980

Chambers, R., 'Sustainable Rural Livelihoods: a key strategy for people, environment and development', in C. Conroy and M. Lipsinoff (eds), *The Greening of Aid*, London, Earthscan Publications, 1989

Chambers, R. and Toulmin, C., 'Farmer-first: achieving sustainable dryland development in Africa', Chapter 3, this volume

Chayanov A. See Thorner, Kerblay and Smith (1966)

Chuta, E. and Liedholm, C., 'Rural non-Farm Employment: a review of the state of the art', Michigan State University, Rural Development Paper No. 4, 1979

Collinson, M., 'The Evaluation of Innovations for Peasant Farming', *East African Journal of Rural Development* 1, 2, July 1968

Conroy, C. and Lipsinoff, M., *The Greening of Aid*, London, Earthscan Publications, 1989

Evans, A., 'Gender Issues in Rural Household Economics', *Discussion Paper No. 254*, Institute of Development Studies, Sussex, 1989

FAO, 'Farm Management and Production Economics Service', *Farm Systems Development: Concept, methods, applications*, FAO, Rome, 1989

Fortes, M., Steel, R. and Ady, P., 'The Ashanti Survey 1945–56: an experiment in social research', *The Geographical Journal* 60, 4–6, 1949

Gilbert, E., Norman, D. and Winch, F., 'Farming Systems Research: a critical appraisal', Michigan State University, East Lansing, Department of Agricultural Economics, MSU Rural Development Papers, No. 6, 1980

Griliches, Z., 'Research Costs and Social Returns', *Journal of Political Economy*, 1958

Guyer, J., 'Intra-household Processes and Farming Systems Research', in J. Moock (ed.), *Understanding Africa's Rural Households and Farming Systems*, Boulder, Col., Westview, 1986

Hunt, D., 'Poverty and Agricultural Development Policy in a Semi-arid Area of Eastern Kenya', in P. O'Keefe and B. Wisner (eds), *Land Use and Development*, International African Institute, 1977

—— *The Impending Crisis in Kenya: the case for land reform*, Aldershot, Gower, 1984

Huss-Ashmore, R., 'Seasonality in Rural Highland Lesotho: method and policy', in AMREF, *A Report on the Regional Workshop on Seasonal Variations in the Provisioning, Nutrition and Health of Rural Families*, Nairobi, AMREF, 1982

Jiggins, J., 'Problems of Understanding and Communication at the Interface of Knowledge Systems', in S. Poats et al. (eds), *Gender Issues in Farming Systems Research and Extension*, Boulder, Col., Westview, 1988

Jones, W., 'Labor and Leisure in Traditional African Societies', *Items* 22, 1, March 1968

Kenya Government, Ministry of Agriculture, Research Division et al. 'Demonstrations of an Interdisciplinary Farming Systems Approach to Planning Adaptive Research Programmes', CIMMYT Eastern Africa Economics Programme, Report No. 1, 1977

Lancaster, K., 'Change and Innovation in the Technology of Consumption', *American Economic Review* 46, 2, 1966

Levi, J., 'Time, Money and Food: Household Economics and African Agriculture', *Africa* 57, 3, 1987

Lipton, M. with Longhurst, R., *New Seeds and Poor People*, London, Unwin Hyman, 1989

Livingstone, I., *Rural Development, Employment and Incomes in Kenya*, ILO, JASPA, Addis Ababa, 1981 (republished with Appendix, Aldershot, Gower, 1986).

Low, A., *Agricultural Development in Southern Africa: Farm Household Economics and the Food Crisis*, London, James Currey, 1986a

—— 'On Farm Research and Household Economics', in J. Moock (ed.), *Understanding Africa's Rural Households and Farming Systems*, Boulder, Col., Westview, 1986b

Matlon, P., 'The Size Distribution, Structure and Determinants of Personal Income among Farmers in the North of Nigeria', D. Phil thesis, Cornell University, 1977

Moock, J., *Understanding Africa's Rural Household's and Farming Systems*, Boulder, Col., Westview, 1986

Mwangi, S., 'Shelter Sure – a roofing tiles project', *Rural Development in Practice* 1, 4, May 1989

Norman, D., 'Initiating Change in Traditional Agriculture', *Proceedings of the Agricultural Society of Nigeria*, 1970

—— 'The Farming Systems Approach: relevance to the small farmer', Michigan State University, East Lansing, Department of Agriculture, MSU Rural Development Papers, No. 5, 1980

—— Simmons, E. and Hays, H., *Farming Systems in the Nigerian Savannah: research and strategies for development*, Boulder, Col., Westview, 1982

O'Keefe, P. and Wisner, B., *Land Use and Development*, International African Institute, 1977

Poats, S., Schmink, M. and Spring, A., *Gender Issues in Farming Systems Research and Extension*, Boulder, Col., Westview, 1988

Sen, A., *Resources, Values and Development*, Oxford, Blackwell, 1984

Shaner, W. Philipp, P. and Schmehl, W. (eds), *Farming Systems Research and Development: Guidelines for developing countries*, Boulder, Col., Westview 1982

Shucksmith, D. et al., 'Pluriactivity, Farm Structures and Rural Change', *Journal of Agricultural Economics* 40, 3, 1989

Singh, I., Squire, L. and Strauss, J., 'A Survey of Agricultural Household Models: recent findings and policy implications', *World Bank Economic Review* 1, 1, 1986a

Diana Hunt

—— *Agricultural Household Models: extensions, applications, and policy*, Baltimore, MD, Johns Hopkins University Press, 1986b
Strauss, J., 'Joint Determination of Food Consumption and Production in Rural Sierra Leone: Estimate of a household-firm model', *Journal of Development Economics* 14, 1984
Thorner D., Kerblay B. and Smith R. (eds), *A. V. Chayanov on the Theory of Peasant Economy*, Illinois, Imoin, 1966
Whitehead, A., 'I'm Hungry Mum: the politics of domestic budgeting', in K. Young, et al. (eds), *Of Marriage and the Market*, London, CSE Books, 1981
Young, K. et al. *Of Marriage and the Market*, CSE Books, 1981

5

Animal Traction: Constraints and Impact among African Households

Paul Starkey

Are farmers better off working with animals? The majority of African farmers still use human energy for crop cultivation, while animal traction remains an innovative technology in many parts of sub-Saharan Africa. It often brings benefits, but these are not necessarily equally distributed within individual households.

At the beginning of the twentieth century European settlers started to use draught animals on their farms in Kenya, Mozambique, South Africa and Zimbabwe, and the technology diffused into some of the surrounding areas. In the period 1900–40 colonial schemes to grow cotton led to the sale of ox-ploughs and the promotion of animal traction among African farmers, including in parts of Cameroon, Chad, Guinea, Mali and Uganda. During the same period, colonial administrations introduced animal traction for the production of groundnuts and food crops in Botswana, Madagascar, Nigeria, Senegal, Sierra Leone and Tanzania.

These early introductions of animal traction were quite localised. In most villages animal traction has only become a known technology within the lifetime of the present village elders and older farmers (those over fifty). In some areas, such as The Gambia, southern Mali, southern Mozambique, central Senegal, northern Nigeria and much of Zimbabwe, the technology started diffusing widely only in the 1950s and 1960s. Meanwhile, in many locations (including Ghana, northern Mozambique, Niger, southern Nigeria, Sierra Leone, most of Sudan, and Zaire) adoption rates are low, despite promotion schemes by past and present governments.

Paul Starkey

Characteristics of Animal Technology in sub–Saharan Africa

In most cases the type of draught animal and the harnessing system that were first introduced have remained in use. Oxen (castrated bulls) are the dominant draught animals in Africa. Cattle have a wide geographic range; cattle herds invariably produce more males than are needed for reproduction; and strong and docile oxen are excellent draught animals. However, in some areas such as northern Nigeria uncastrated males (bulls) are used for work, while in Sine Saloun (Senegal) cows (females) are increasingly being worked. In general, the oxen used are those found locally. However, cattle production in Africa is often stratified, with the offtake of the large reproductive herds based in arid range lands being progressively sold to markets in neighbouring areas, and beyond. Thus cattle in local markets may well have come from far away, even from other countries. However deliberate attempts to introduce exotic species or breeds for draught power have had a negligible impact, mainly due to management and health problems (Starkey,1985).

Most oxen are humped 'zebu' or 'sanga' cattle of various breeds, but in the tsetse-infested areas of west and central Africa humpless, trypano-tolerant taurine breeds are used. The 'best' males, those that are large and strong, are selected for work. These are therefore castrated and so cannot be used for breeding purposes. Thus the breeding bulls may be genetically inferior in body size and conformation, giving rise for example, to worries over the 'shrinking Mashona beast', as in Zimbabwe (Tembo, 1989). Similar trends associated with castration to produce draught animals have been observe¹ in southeast Asia (Momongan et al., 1989); but this is less likely to occur in Sahelian countries wherever the large reproductive herds are not controlled by crop farmers.

Most cattle are yoked in pairs with withers yokes; but horn/head yokes are often used with the humpless taurine cattle of West Africa. Multiple teams of 6–12 cattle are commonly used in Botswana and teams of 4–6 cattle are sometimes used in the other countries of southern Africa. However, multiple teams are rarely used in Ethiopia, West or Central Africa. Single oxen are not generally used in Africa, except in Ethiopia; a few are used for pulling carts in Zanzibar.

Horses have a very limited geographical range in Africa. They are not hardy, but are often expensive, as a result of their high prestige value, their suitability for transport and their relatively low reproductive efficiency and survival rate. They are thus seldom used for agriculture, with the notable exception of west-central Senegal, where they are widely used to pull cultivation tines, seeders, weeders and groundnut lifters.

Donkeys have a greater range than horses, and generally have better rates of reproduction and survival. They are well suited for pack transport and for pulling carts in relatively flat areas: however, they do not have great tractive power, and so their use for pulling ploughs is very limited. Since in most African countries the implements first introduced for use with draught animals were ploughs and heavy cultivators designed for strong animals, donkeys were seldom considered for crop cultivation. However, in a few southern African countries large teams of donkeys or mules were sometimes employed. With the development of implements and techniques for low-draught tine cultivation, together with changing ecological conditions, donkeys are increasingly employed for cultivation in West Africa.

Constraints on the Use of Draught Animals

There are some areas where animal traction has been rapidly adopted, and other areas where adoption has been almost non-existent. Many reasons for low adoption rates have been cited, and these are often very location-specific. Pingali, Bigot, and Binswanger (1987) consider that the overriding issue has been the intensity of farming: animal-powered cultivation only becomes viable when the pressures of population on land are sufficient to justify destumping. However, this does not explain why apparently similar areas, either within countries or in neighbouring countries, have very different adoption rates. Low adoption rates have been variously ascribed to inadequate capital/credit, lack of animals, animal disease, lack of suitable implements, social prejudices or lack of knowledge. Sometimes it is difficult to identify clear causes. In one survey along the 'plough line' zone of Nigeria between animal-traction users (to the north) and non-users (to the south) no single reason stood out for the relatively clear demarcation line: implicated factors included animal disease, socio-ethnic differences, extension advice, availability of tractor-hire schemes and implement availability (Blench, unpublished). But elsewhere in Africa there have been cases where providing one particular service or input (credit, extension, veterinary skills, implements) has led to the adoption of animal traction; such examples suggest that there can be specific limiting factors, which stop animal traction from diffusing. Yet within the areas benefiting from such services there have almost always been localities where there has been no take-off.

To give some examples: poor animal condition due to inadequate nutrition at the end of the dry season is often cited as a constraint on animal use for cultivation in west Africa. However, in parts of Mali, Senegal and The Gambia (where such a constraint is widely acknowledged) many farmers stock groundnut hay for use by transport animals. In Zaire two projects had problems of poor infrastructure and farmer unfamiliarity with draught animal technology. In one area farmers started to benefit from increasing maize prices as traders carried grain to a growing town, and in this area animal traction adoption was higher. There was no suggestion that the social or technical constraints were any less, but in one area economic profitability was higher. To take another example, much attention has been paid to the design of animal-drawn carts in Africa, particularly to ways of overcoming the serious constraint of punctures of pneumatic tyres. Yet there is plenty of evidence to suggest that punctures can be repaired at village level whenever it is economically (a bush taxi) or socially (a prestige moped) appropriate. In short, punctures are a genuine technical constraint, but one that can be overcome. Other examples could be cited where constrainst have been overcome as farmers (or itinerant traders) have travelled on their own initiative to purchase animals, implements or veterinary drugs, or to learn from farmers already using draught animals. In such cases constraints have been overcome without public sector 'project' initiatives.

The importance of economic profitability has been stressed to illustrate the interaction of limiting factors. Even constraints that may seem overriding (such as the absence of suitable animals in areas infested with the disease trypanosomiasis) can often be overcome (e.g. by buying in animals and providing chemical prophylaxis) if the benefits justify the costs.

In early stages of adoption, animal traction increases the risk of capital loss. The risk of animal mortality is particularly important for farmers who were not animal owners, and who therefore purchase animals using savings or credit. Such people are often unused to animal husbandry and yet have committed themselves to a major investment. Animal mortality rates in some areas of introduction have been high, with figures in excess of 25 per cent reported for particular schemes in Burkina Faso, Cameroon, Malawi and Sierra Leone. In areas where draught animals are in regular use, wastage rates of 3–10 per cent can be experienced through disease, accidents (e.g. broken limbs, eye damage, poison, bloat, snakes, lightning) or theft. Farmers sometimes try to reduce risk through their choice of animal. In The Gambia donkeys have been seen to have a much lower risk of theft than cattle (donkey meat has no value there), even though morality rates are greater. In western Sudan, promotion of the use of camels for draught purposes was suspended after theft reached unacceptable levels, and attention turned to using draught cattle and donkeys. Some credit schemes (including some in Burkina Faso and Togo) have had animal insurance built into the cost of credit, but verification of insurance claims has often proved difficult.

Economic Impact

It is difficult to assess the economic impact of animal traction, for it cannot easily be assessed in isolation from numerous other inter-acting elements of farming systems. To ensure observed trends are not spurious, economic information has to be collected over a period of time (several seasons if possible) and there are almost always 'exceptional' social, economic or environmental conditions to contend with, large variations between farmers and years, changing economic conditions and government interventions and other confounding factors (Starkey, 1989). It is also difficult to ascribe opportunity costs to the labour required by, or saved with, animal traction, particularly when it causes shifts in the time and category of labour. Quite small amounts of time saved by adults during crucial labour-bottleneck cultivation periods may have to be 'paid for' by much longer periods of child labour, most of which will be required during slacker periods of the year. Reddy (1988) also notes that economists have tended to underestimate the sever-ity of initial cash-flow problems. Starkey (1988b) notes that many sets of costings were produced to illustrate the profitability of adopting animal-drawn, wheeled toolcarriers, but they convinced donor agencies rather than farmers. It seems to be increasingly realised that in the recent past a large discrepancy has often existed between the economic perceptions of farmers and those of conven-tional project economists (Reddy, 1988; Starkey, 1989; Bordet,1990).

Animal traction may lead to changes in the crop mix, and thus may have differential effects on crop production. Animal traction has often been promoted in Africa for mono-cropping in areas where inter-cropping was traditional, and it has sometimes been assumed that it leads to increased production of cash-crops, such as cotton, to the detriment of food-crops, such as maize. However, some surveys have found no marked differences (Barratt et al., 1982; McIntire, 1983; Panin, 1986; Francis, 1988). Increased production of cash-crops does not necessarily imply lower production of food crops. In the cotton zones in West Africa, where animal traction has been successfully promoted by cotton companies, food grain production has increased too. It is thought that grain production benefits not only from animal power for cultivation, but also from the residual effects of fertiliser applied to cotton crops (Deveze and Levaray, 1988). Nevertheless should a major change in crop mix be associated with animal traction, this could well have significant effects on food production and the local economy, with different effects on the various members of farm households.

Animal traction is often associated with higher crop yields per farm and farmer (and sometimes per hectare) than those obtained by hoe-farming. However, there is seldom a direct effect of the tillage *method* on land yields, although the *timeliness* of ploughing, seeding and weeding may improve with animal traction, leading to yield increases. Line planting, followed by early or regular inter-row weeding with draught animals, may also improve yield per unit of labour and area. Working with farmers in Ghana, Panin (1986) found that a significant increase in yield per hectare of maize, millet, groundnuts and beans was associated with ridge formation using oxen. Research trials in Ethiopian black soils (Vertisols) have demonstrated that broad beds or ridges made with oxen produce more reliable grain yields and significantly higher yields in years with poor rainfall distribution (Jutzi, Anderson and Astatke, 1988). Meanwhile, some yield increases may well be due to factors associated with, but not caused by, the working animals. For example, farmers that have adopted work oxen may be more likely to use fertilisers than other farmers.

When animal traction technology is new, extensification rather than intensification is the norm. Overall crop output generally increases, mainly due to the area expansion. Men and boys usually train the animals, work with them and herd them. These people have the initial problems associated with first use of animals and area expansion.

In some communities men consider it appropriate to cultivate land for crops usually grown by women: in others they do not. Women and children often have the task of weeding and harvesting, and their work may be increased if cultivated areas are expanded. Children often tend draught animals; because of this their educational prospects can suffer, either due to limited school attendance or to fatigue when school is combined with looking after animals. In one small survey in Sierra Leone, it was found that children of draught animal owners were less likely to attend primary school (Allagnat and Koroma, 1984).

In sub-Saharan Africa most of the direct economic costs and benefits of animal traction (relevant capital and recurrent expenditure, the cost of credit and the income from hiring and harvests) are controlled by males. There are examples of women owning draught animals and being given credit through banks or projects, but these are exceptions. While in most African countries it is common to see women cultivating fields with hoes, it is rare to see women ploughing with animals. In West Africa, women may well lead the animals, but in almost all cases a man handles the plough. Women can, however, be observed ploughing in parts of Malawi and Tanzania, and women have been trained to plough in certain small projects in Cameroon, Sierra Leone and Zaire. When women have access to the use of draught animals it is often through informal exchange or hire arrangements, involving the services of male handlers as well as the animals.

In several countries in Africa, animal traction has been introduced through communal ownership, often encouraged by government or aid agencies. While there have been examples of successful village associations for animal traction, many have experienced major social and organisational problems associated with conflicting interests for access during the crucial working hours and responsibility for maintaining the animals at other times. With individual ownership it is clear who is responsible for both the costs and the benefits of animal management. One of the costs is grazing supervision, and if this is not carried out effectively the animal may suffer from insufficient food, accident or theft: alternatively, growing crops can be eaten, causing much social conflict and expense. In one survey in Sierra Leone a quarter of ox-owners reported that they have to pay out significant sums in compensation as a result of the alleged damage caused by their work oxen (Corbel, 1988).

The use of animals for transport also provides numerous opportunities for social and economic benefits including:

1 Reduced porterage, i.e. the direct carrying of goods by men or women by head-load;
2 Quicker and more efficient evacuation of harvested produce involving less harvesting losses;
3 Improved access to markets;
4 Greater ease of utilising crop residues, composts, manures and purchased farm inputs.

About 600,000 animal-drawn carts are in use in sub-Saharan Africa and about 10 per cent of African farmers who own draught animals have a cart. The number of carts in use has increased greatly in recent years and the importance of carts to the agricultural sector is much greater than absolute numbers imply, since carts (unlike ploughs) are used throughout the year. Indeed, in parts of central Kenya where a combination of factors – the extent of tree-cropping, population and land-use intensity, and steep slopes – jointly militate against the use of animal adoption for farming, donkey power has become widely used for local transport.

The pattern of adoption of cart technology in sub-Saharan Africa is very uneven. Recent adoption of animal-drawn carts has been particularly noticeable in Senegal and Mali, where relatively high-cost, steel-framed carts, fitted with roller-bearings and pneumatic tyres, have proved very popular. Attempts to develop cheaper carts in Southern and Eastern Africa have seldom been very successful. In Ethiopia, a country with numerous draught and pack animals, there is some horse-drawn urban transport, but very few carts are seen in rural areas. In Madagascar, on the other hand, there are few pack animals but wooden carts with large, spoked wheels play an important role in the rural economy, where frequently used tracks scar the highlands.

While carts are relatively complex and expensive, simple wooden sledges can be made by selecting a naturally occurring fork of a tree or by joining two wooden beams in the form of a V. Such sledges are used in several areas of Eastern and Southern Africa, and in Madagascar. They have the advantages of being cheap and simple to make and maintain, and they can be used on tracks unsuitable for carts. However, they tend to accelerate erosion by leaving rutted tracks, often only passable by other sledges, which become water courses during heavy rains. In several areas of Southern Africa, including Lesotho and Zimbabwe, the dangers caused to the environment by sledges have led them to be officially discouraged and even banned.

Pack transport using donkeys, horses and mules is well established in Ethiopia, and camels are used for packing in the countries bordering Sahara. Elsewhere, pack animals are only commonly used by traditional pastoralists and transhumant groups, and there has been little recent adoption of this form of transport. This may be linked to the superiority of carts for larger quantities of goods and the role of women as transporters of small quantities.

Ecological Effects

Animal traction is associated with increased farming intensity, deforestation and permanent cultivation, although it is not usually the primary cause of these. Population pressure is causing African farming systems to evolve from shifting forest-fallow cultivation to annual cropping in destumped fields (Pingali, Bigot and Binswanger, 1987). Animal traction may be a part of this progression with farmers destumping their land for ploughing as farming with short bush-fallows starts to give low returns to labour, and the cost of land preparation and weed control using traditional techniques become excessive.

Permanent cultivation in the absence of soil conservation techniques and the replacement of nutrients can lead to increased erosion. Thus animal traction too may be associated with increased erosion, but not in a cause-and-effect relationship: poor farming practices can lead to erosion whether human labour, animals or tractors are used. In southern Mali some heavily eroded fields have never been cultivated with draught animals, but because most of the land is now tilled with oxen, most erosion is on animal-tilled fields.

Permanent cropping and mono-cropping lead to reduced ecological diversity, and fewer species of native plants and animals, whether they be trees, shrubs, medicinal plants, wild mammals or insects. This depletion of the environment and reduced ecological stability is also associated with the adoption of animal traction technology, but again the association is not cause-and-effect, but linked effects of intensified agriculture.

Increases in the local population of large animals can lead to pasture degradation, and one of the reasons for retaining animals may be for work. Where watering places are few, such as in areas of the Sahel, in Botswana and in southern Mozambique, regular trampling along paths in the vicinity of water holes can cause serious erosion problems. Again, however, the association with animal traction is indirect, and there is no suggestion that animal traction *per se* causes pasture degradation and erosion; indeed, it is often argued that animal traction encourages crop–livestock integration.

The traditional separation of livestock rearing and crop production found in some African countries can become socially divisive and environmentally unacceptable as population and land pressures increase. Animal traction allows nutrients to be recycled and soil fertility to be maintained through the use of animal dung, as well as green manure and composting techniques. Farm carts are particularly useful for transporting residues and fertilisers. Farmers learn animal husbandry techniques when they start working with draught animals; and attitudes and skills learned in this way may be applied to other livestock enterprises.

Tractors as an Alternative

The relative success of animal traction adoption has often been contrasted with the failure of tractorisation schemes. During the 1960s and 1970s most countries in sub-Saharan Africa attempted to bring tractors to their smallholder farmers, often through tractor-hire schemes. A few private-sector initiatives proved viable in countries with large-farm sectors, such as Kenya, South Africa and Zimbabwe. Other private sector tractor schemes developed in the countries bordering on South Africa, such as Botswana, Lesotho and Swaziland, where low-cost, second-hand tractors were available and where land preparation costs were often financed or subsidised by remittances from migrant workers. However, most government-run schemes proved economically unsustainable.

It was the failure of many public sector tractorisation schemes that led the World Bank to carry out a two-year research study into the evolution of agricultural mechanisation in sub-Saharan Africa. This concluded that most African farming systems will go through stages of gradual intensification, and almost invariably animal traction technology will become economically viable before tractorisation (Pingali, Bigot and Binswanger, 1987). However, in some areas, which support a commercial farming sector, a dual system can be found in which tractors are kept for heavy work, and animals for the lighter, more time-consuming agricultural operations.

Conclusion

In conclusion, animal traction is spreading in sub-Saharan Africa, and in many instances is offering rising populations an ecologically sustainable means of increasing agricultural production, reducing human drudgery and improving the quality of rural life. The diffusion of technology has come through formal programmes and informal diffusion, and has been fastest when profitable markets have been available to justify the investment. The provision of training, implement supply, credit and other back-up services by publicly funded agencies or private companies can aid adoption, and the process often seems to accelerate once the critical level at which small traders and the informal traditional sector begin to offer their own specialised support services. The change to animal traction necessitates the surmounting of many constraints, but problems can generally be solved where the technology is clearly profitable.

Author's Note

As chapters 7, 8 and 10 in this volume illustrate, animal traction has proved viable over a wide spectrum of degrees of incorporation into the broader market economy. It is likely to play a positive and increasing role in African agriculture well into the twenty-first century.

References

Allagnat, P. and Koroma, B., 'Socio-economic Survey of the Use of Ox Traction in the Mabole Valley, Bombali District', Sierra Leone Work Oxen Project and Association Française des Volontaires du Progrès, Freetown , Sierra Leone, 1984

Barratt, V., Lassiter, G., Wilcock, D., Baker, D. and Crawford, E., 'Animal Traction in Eastern Upper Volta: a technical, economic and institutional analysis', *International Development Paper 4*, Department of Agricultural Economics, Michigan State University, East Lansing, Michigan, 1982

Blench, R. M., 'Social Determinants of Animal Traction in Central Nigeria', Agricultural Research Unit, The World Bank, Washington DC (unpublished)

Bordet, D., 'La traction animale dans les systèmes de production: effets dynamiques', in P. Starkey and A. Faye (eds), *Animal Traction for Agricultural Development*, Proceedings of workshop held 7–12 July 1988, Saly, Senegal. International Livestock Centre for Africa (ILCA), Addis Ababa, 1990 (in press)

Corbel, H., 'The Economics of Animal Power in Koinadugu District,

Sierra Leone: a case study of the work oxen introduction and credit programme', in P. H. Starkey and F. Ndiamé (eds), *Animal Power in Farmer Systems*, pp. 299-310. Proceedings of networkshop held 17–26 September 1986, in Freetown, Sierra Leone. Vieweg for German Appropriate Technology Exchange, GTZ, Eschborn, Germany, 1988

Deveze, J. C. and Levaray, G., 'Questions sur l'évolution et les perpectives de la culture du coton en zone de savane' (Afrique de l'Ouest et du Centre), in G. Raymond (ed.), *Economie rurale en zone de savane*, pp. 116–29, Actes due Vlle Seminaire d'économie et sociologies rurales, 15–19 September 1986, Montpellier. Centre de Coopération Internationale en Recherche Agronomique pour le Développement (CIRAD), Montpellier, 1988

Francis, P. A., 'Ox Draught Power and Agricultural Transformations in Northern Zambia', *Agricultural Systems* 27 (1), 1988, pp. 35–49 Gboku, M., 'Farmer Social Variables Influencing the Adoption of Agricultural Innovations in Sierra Leone', in P. H. Starkey and F. Ndiamé (eds), *Animal Power in Farming Systems,* pp. 311–19, Proceedings of networkshop held 17–26 September 1986 in Freetown, Sierra, Leone. Vieweg for German Appropriate Technology Exchange, GTZ, Eschborn, 1988

Gryseels, G., Abiye Astatke, Anderson, F. M. and Getachew Assemenew, 'The Use of Single Oxen for Crop Cultivation in Ethiopia', *ILCA Bulletin* 18, 1984, pp. 20–5

Jahnke, H. E., *Livestock Production Systems and Livestock Development in Tropical Africa*, Kieler Wissenschaftsverlag Vauk, 1982

Jutzi, S. and Goe, M. R., 'Draught Animal Technologies to Intensify Smallholder Crop-livestock Farming in the Ethiopian Highlands', Paper prepared for FAO World Animal Review, International Livestock Centre for Africa (ILCA), Addis Ababa, Ethiopia (unpublished), 1987

—— Anderson, F. M. and Abiye Astatke, 'Low-cost Modifications of the Traditional Ethiopian Tine Plow for Land-shaping and Surface Drainage on Heavy Clay Soils: preliminary results from on-farm varification', in P. Starkey and F. Ndiamé (eds), *Animal Power in Farming Systems*, pp. 127–38, Proceedings of workshop held 17–26 September 1986, Freetown, Sierra Leone. Vieweg for German Appropriate Technology Exchange (GATE), GTZ, Eschborn, 1988

Kanu, B. H., 'Animal Traction Development Strategies in Sierra Leone', in P. H. Starkey and F. Ndiamé (eds), *Animal Power in Farming Systems*, pp. 277–87, Proceedings of networkshop held 17–26 September 1986 in Freetown, Sierra Leone. Vieweg for German Appropriate Technology Exchange, GTZ, Eschborn, 1988

Lekezime, P., 'Mechanical Weeding with Animal Traction: some prerequisites', in P. H. Starkey and F. Ndiamé (eds), *Animal Power in Farming Systems*, pp. 350–2, Proceedings of networkshop held 17–26 September 1986 in Freetown, Sierra Leone. Vieweg for German Appropriate Technology Exchange, GTZ, Eschborn, 1988

Lhoste, P., 'La gestion de la carrière des bovins de trait: élément important de la rentabilité de l'utilisation de la traction bovine', in P. Starkey and A. Faye (eds), *Animal Traction for Agricultural Development*, Proceedings of workshop held 7–12 July 1988, Saly, Senegal. International Livestock Centre for Africa (ILCA), Addis Ababa, Ethiopia (in press)

Loewen-Rudgers, L., Rempel, E., Harder, J. and Klassen Harder, K., 'Constraints to the Adoption of Animal Traction Weeding Technology in the Mbeya Region of Tanzania', in P. Starkey and A. Faye (eds), *Animal Traction for Agricultural Development*, Proceedings of workshop held 7–12 July 1988, Saly, Senegal. International Livestock Centre for Africa (ILCA), Addis Ababa, Ethiopia (in press)

McIntire, J., 'Two Aspects of Farming in SAT Upper Volta: animal traction and mixed cropping', *Progress Report* 7, ICRISAT Economics Program, Ouagadougou, Burkina Faso, 1983

Momongan, V. G., Parker, B. A., Santos, E. B. de los and Ranjhan, S.K., 'Breeding Programs for Improved Draught Animal Power: crossbreeding of buffaloes', in D. Hoffman, J. Mari and R. J. Petheram (eds), *Draught Animals in Rural Development*, pp. 190–4, Proceedings of an international research symposium held at Cipanas, Indonesia, 3–7 July 1989. ACIAR Proceedings Series No. 27, Australian Centre for international Agricultural Research, Canberra, 1989

Panin, A., 'A Comparative Socio-economic Analysis of Hoe and Bullock Farming Systems in Northern Ghana', PhD thesis, University of Guettingen, 1986

Pingali, P., Bigot, Y. and Binswanger, H., *Agricultural Mechanisation and the Evolution of Farming Systems in Sub-Saharan Africa*, World Bank, Washington, in association with Johns Hopkins University Press, Baltimore, MA 1987

Reddy, S. K., 'Use of Animal Power in West African Farming Systems: research level problems and implications for research – perspectives from Mali', in P. Starkey and F. Ndiamé (eds), *Animal Power in Farming Systems*, pp. 182–90. Proceedings of networkshop held 19–26 September 1986, Freetown, Sierra Leone. German Appropriate Technology Exchange, GTZ, Eschborn, 1988

Starkey, P. H., *Farming with Work Oxen in Sierra Leone*, Ministry of Agriculture, Freetown, Sierra Leone, 1981

——, 'Genetic Requirements for Draught Cattle: experience in Africa', in J. W. Copland (ed.), *Draught Animal Power for Production*, Proceedings of an international workshop held at James Cook University, Townsville, Qld, Australia, 10–16 July 1985, ACIAR Proceedings Series 10, Australian Centre for International Agricultural Research, Canberra, 1985

——, 'Brief Donkey Work', *Ceres* 20, 6, 1987, pp. 37–40

——, *Animal Traction Directory: Africa*, Vieweg for German Appropriate Technology Exchange, GTZ, Eschborn, 1988a

——, *Perfected yet Rejected: Animal-drawn Wheeled Toolcarriers*, Vieweg for

Germany Appropriate Technology Exchange, GTZ, Eschborn, 1988b

——, 'The Introduction, Intensification and Diversification of the Use of Animal Power in West African Farming Systems: implications at farm level', in P. H. Starkey and F. Ndiamé (eds), *Animal Power in Farming Systems*, pp. 97–115, Proceedings of networkshop held 17–26 September 1986 in Freetown, Sierra Leone. Vieweg for German Appropriate Technology Exchange, GTZ, Eschborn, 1988c

——, 'Animal Traction Research in Southern Mali', Consultancy report for Division de Recherche sur les Systèmes de Production Rurale (DRSPR), Sikasso, Mali (unpublished), 1988d

——, 'An Overview of Farming Systems Research Relating to Draft Animal Power', in D. Hoffman, J. Mari and R. J. Petheram (eds), *Draught Animals in Rural Development*, Proceedings of an international research symposium held at Cipanas, Indonesia, 3–7 July 1989, ACIAR Proceedings Series No. 27, Australian Centre for International Agricultural Research, Canberra, 1989

—— and Ndiamé, F. (eds), *Animal Power in Farming Systems*, Proceedings of networkshop, held 17–26 September 1986 in Freetown, Sierra Leone. Vieweg for German Appropriate Technology Exchange, GTZ, Eschborn, 1988

Tembo, S., 'Draught Animal Power Research in Zimbabwe: current constraints and research opportunities', in D. Hoffman, J. Mari and R. J. Petheram (eds), *Draught Animals in Rural Development*, pp. 61–8, Proceedings of an international research symposium held at Cipanas, Indonesia, 3–7 July 1989, ACIAR Proceedings Series No. 27, Australian Centre for International Agricultural Research, Canberra, 1989

Westneat, A. S., Klutse, A. and Amegbeto, K. N., 'Features of Animal Traction Adoption in Togo', in P. H. Starkey and F. Ndiamé (eds), *Animal Power in Farming Systems*, pp. 331–9, Proceedings of networkshop held 17–26 September 1986, in Freetown, Sierra Leone. Vieweg for German Appropriate Technology Exchange, GTZ, Eschborn, 1988

White S., 'The Bauchi State Agricultural Development Project', *Draught Animal News* (Centre for Tropical Veterinary Medicine, Edinburgh) 4, 1985, pp. 17–20

Part II

Case Studies

6

Autarky and Technical Change in Rice Production in Guinea Bissau: on the Importance of Commoditisation and Decommoditisation as Interrelated Processes

Jan Douwe van der Ploeg

Autarky is often considered to be one of the major obstacles to rural change. In this chapter I will argue that it is not. When deconstructing the global notion of autarky, several forms of relative autonomy of farming *vis-à-vis* the market can be conceptualised. These are not limited to those areas normally associated with agricultural backwardness. On the contrary, they are as present in European as in African agriculture, and more often than not they have formed and still form the basis for impressive processes of agricultural growth (van der Ploeg, 1985, 1986a, b). The main problem is that technical innovations are designed in ways that do not necessarily correspond to prevailing patterns of autarky. Their impact is therefore limited or even counterproductive.

Guinea Bissau has been linked to world market circuits for centuries. Both the mechanisms involved and the intensity of the specific interrelations have changed frequently over time and often quite abruptly. However, the latter does not exclude, from an analytical point of view, Guinea Bissau being characterised as having an economy in which the generalised circulation of commodities is dominant (Bernstein, 1985). No one, not even in the most remote corners of the country, can reproduce his or her life without entering into commodity circuits. Autarky is thus in its classical sense, as illustrated by Robinson Crusoe on his island, a void category. And always was, as Bloch, the French agricultural historian, demonstrates in his 'Economie-nature ou économie-argent' and highlights in the subtitle: 'un pseudo-dilemme' (Bloch, 1939).

The linkages between farming and markets can be analysed on two levels: first, the input side of farming, and second, the 'output side'. On the input side, the labour force needed to produce the labour objects (soil, water, seeds, etc.) and the instruments (or means of labour) may all be acquired in the respective markets and will then enter the labour process as commodities. However, they can be produced and reproduced equally well within the farm and farming family. They will then enter the labour process as particular values having a precise social value: they enable the farming family (or community, or whatever the relevant social unit might be) to produce a living as well as a future. The specific nature of these will depend on the prevailing social relations of production.

On the output side a similar range of possibilities can be discerned. Farm products can be sold, (i.e. be turned into commodities); they can be used for internal consumption or for future cycles of production; or they can be exchanged through socially regulated circuits.

In simple commodity production (i.e. labour processes devoted to the production of value, not surplus value), which is the dominant form of production in agriculture – in Guinea Bissau this is the only form – there will always be a specific balance between commodity and non-commodity relations. Within the theoretical space, which ranges from farm units merely linked to markets through the selling of some of their end-products, to farms whose operation and development is based on a 'complete circulation of commodities', all kinds of analytical as well as empirical diversity can be localised. The particular balance between commodity and non-commodity relations, however, is not incidental. It is the outcome of complex and often interrelated processes of commoditisation and decommoditisation. This implies that:

1 commodity circuits and commodity systems should not be conceptualised as kinds of mechanical or functional units smoothly following the 'logic of the markets', but as highly complex configurations, generating all kinds of internal contradictions;

2 consequently, commodity and non-commodity circuits cannot be simply juxtaposed or presented in a hierarchical order in which the one is subsumed or simply functional to the other: there are specific forms of commoditisation that allow for specific forms of decommoditisation. Frequently, the *raison*

d'être of specific forms of decommoditisation is that they allow for specific forms of what appear to be the opposite, i.e. specific forms of commoditisation.

In short, commodity and non-commodity circuits, commoditisation and decommoditisation should not be conceptualised as mutually exclusive in time and space. Rather, they should be seen as dialectical unities, and hence as entities that not only go together, reproducing each other continuously, but also as being subjected to the continuous and often conflicting strivings of the different actors involved (Long, 1985; Long and van der Ploeg, 1988). We will touch on this last element by describing the shift from north to south of the Balanta, the main rice-growers in Guinea Bissau. The Balanta are also one of the most important ethnic groups in Guinea Bissau, representing some 30 per cent of the total population. Historically, they represented the main social force in the anti-colonial struggle.

Rice and Food Security: a Tidal Wave

Rice is the main food-crop in Guinea Bissau; it is also, according to the political and economic circumstances, one of the main export crops. Figure 6.1 summarises the trends in rice imports and exports.

Available data indicate that between 1880 and 1890 rice was exported, after which rice production diminished, making imports necessary. It was only at the beginning of the 1930s that this negative trend made way for exports, with a peak being reached in 1947. Since then, a consistent downward trend can be seen. The 1970s was characterised by chronic shortages and high imports. Only very recently (1987 and 1988) does there seem to be have been a slight shift upwards (Ribeiro, 1988). The tidal wave of imports and exports, as presented in Figure 6.1, is, of course, strongly linked to the rise (or fall) of rice production, as well as to the question of food security. Available reports from the 1920s frequently refer to hunger and to the scarcity of rice. This situation was chronic in the 1970s and into the 1980s. Thus, food shortages are not just a problem of our times: they recur over and over again. At the same time, it should be stressed that the capacity to produce enough, even allowing for considerable exports, is not simply a phenomenon of our times. On the contrary, abundant exports made way for a chronic dependency on food imports. To put this

Figure 6.1 Historical trends in imports and exports of rice in Guinea Bissau

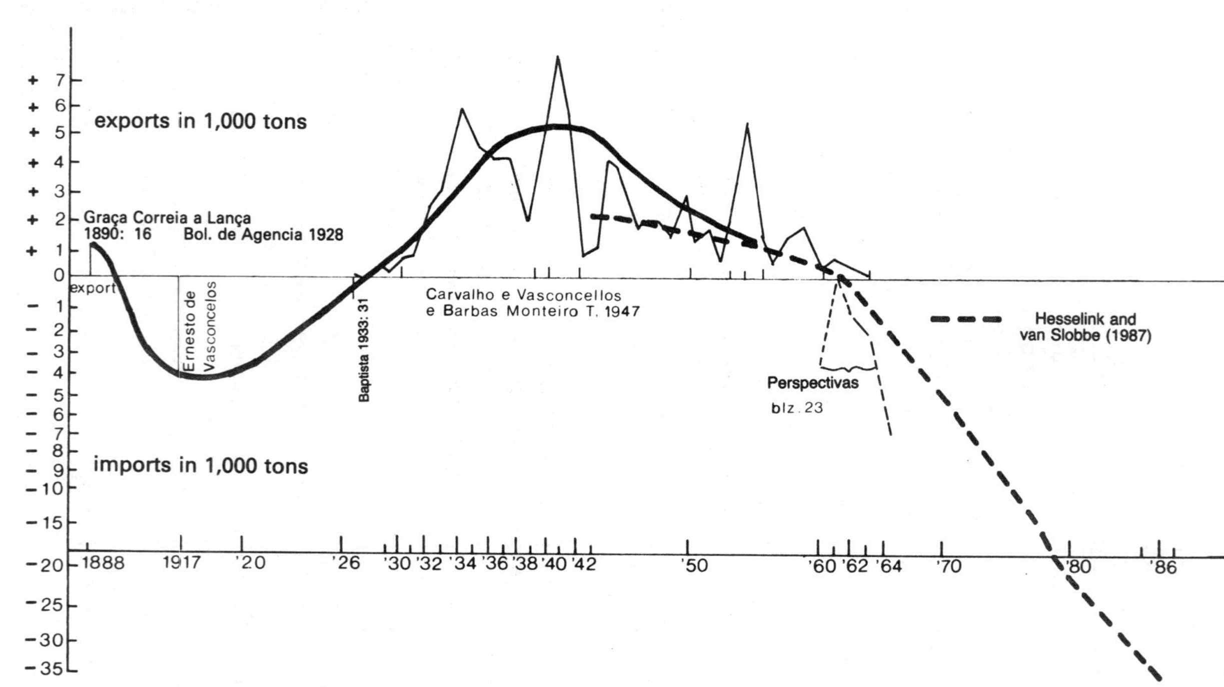

into proper perspective, rice growers in Guinea Bissau only sell rice (thus making it available for export) when consumption at household level is assured. Thus, Figure 6.1 can also be interpreted as indirectly reflecting the ever-changing food requirements of the population.

North and South: Commoditisation and Decommoditisation

The tidal wave represented in Figure 6.1 can be broken down into two separate, albeit combined, historical movements: one in the north of the country, the original home country of the main rice growers, i.e. the Balanta; the other in the south, the virgin land to which the Balanta moved in order to reconstruct their way of life and working once their original homeland had been destroyed (see for a detailed description van der Ploeg and van Slobbe, 1988, 1989). As indicated in Figure 6.2, the tidal wave of rice production (and consequently, of rice imports and exports), can be broken down into two geographically discrete processes. History first witnesses the rise and subsequent fall of the northern rice system, and then the rise and again the fall of the southern rice system. Commoditisation and decommoditisation play a crucial role in these processes.

Rice traditionally was, and continues to be cultivated in *bolanhas*, tropical rice polders, which require a massive, well-organised and stable labour input. If these conditions are fulfilled, production and productivity are quite high; otherwise a considerable and irreversible fall occurs. The beginning of the twentieth century was characterised by just such a fall, when prolonged 'civilising' wars subordinated the Balanta to colonial rule. The wars were followed by forced labour, the effects of which were as devastating as the wars. During the same period the cultivation of groundnuts expanded relatively quickly. To a considerable degree, this was a case of self-commoditisation. Confronted with lower yields in their *bolanhas*, the Balanta tried to redress the problem by cultivating groundnuts. While rice was mainly distributed according to socially regulated exchange rules (only a part of the surplus was sold, as was the case at the end of the nineteenth century), groundnuts always figured as a commodity. As such they introduced a new organising principle, i.e. the maximisation of monetary income per male household head. This same principle filtered down into the organisation of rice production, thus causing a far-reaching

Figure 6.2   Rise, decline and fall of rice-producing systems

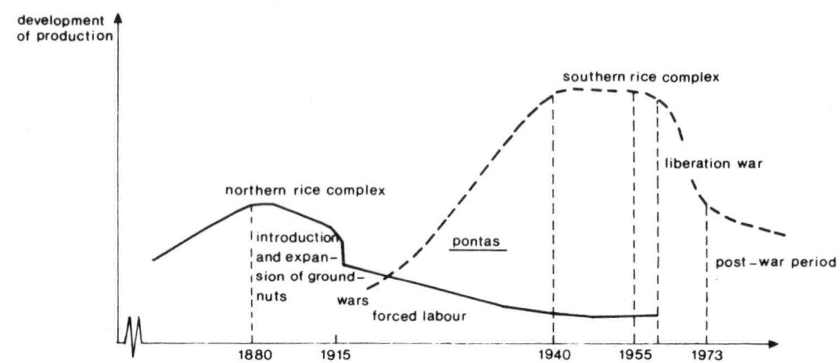

restructuring of the labour process. A persistent extensification of rice production was one of the outcomes. While rice yields fluctuated between 2,000 and 3,000 kg/ha at the beginning of the twentieth century, the mean yield in the north in 1954, was as low as 760 kg/ha.

Commoditisation of the northern rural economy implied a basic transformation of rice: from being initially a commodity, at least partly , it was transformed into a mere subsistence product. At the same time a surplus abundance was turned into a chronic scarcity.

For these and other reasons, an increasing number of Balanta people sought a new life in the (hitherto sparsely populated) south. This migration contrasted sharply with earlier ones: it was not temporary, but permanent, symbolised by the fact that the Balanta gods came to the south as well. Neither was this migration inspired and conditioned by a certain 'logic of the market', as happened with the continuous migration to virgin soils for the cultivation of groundnuts. Rather, it was a migration away from the logic of the market, as established in the north. This should not be seen to imply that the shift southward is to be considered as a simple jump into a natural economy. On the contrary, the particular autarky that arose in the southern system is to be seen as the outcome of a slow but persistent process of emancipation from commodity circuits, which were at first quite important.

Initially, it was only possible for the Balanta to go south by subordinating themselves to the so-called *ponta* system. If they did not, the colonial authorities would not give them permission to migrate. A *ponta* then represented the claim and authority of Portuguese traders over certain territories, in which local farmers could only gain access to the land when accepting the conditions

98

imposed by these traders. In these *pontas*, foodstuffs (including rice), rice seed and even the right to till the soil were all turned into commodities, to be repaid with the harvest, which was thus also turned into a commodity. But, once established, the Balanta liberated themselves from the system, mainly through the persistent use of 'weapons of the weak' (such as failure to sell to the Portuguese traders), thereby precipitating a profound crisis for these *pontas*. Then there was a considerable demand for wage-labour in nearby Guinea Conakry. The Portuguese authorities even feared that this might provoke another shift further to the south. But the Balanta did not go. They did not turn their labour force into a commodity. Arguably, one of the most convincing facts is that initially the Balanta constructed *bolanhas* as commodities: they constructed them simply in order to sell them to people belonging to other ethnic groups. But this practice was soon abandoned. Another important element relating to this conscious decommoditisation is the fact that although the south is ecologically suitable for groundnut cultivation (and other ethnic groups grow groundnuts on a sizeable scale), the Balanta never cultivated them in a quantity beyond what they needed for family consumption. The shift towards the south, therefore, was a definite rejection of groundnut cultivation and its associated commodity circuits and principles.

Why did the Balanta turn southward, while other ethnic groups, who were similarly affected by Portuguese rule and the shift to groundnuts, did not? As will be outlined in the following sections, the social relations of production as embedded in family and village relations, in gender relations as well as in relations between age-groups, do not fit a far-reaching commoditisation of their economy. In particular, the conversion of rice into a mere commodity would run counter to the internal rationality and dynamics of Balanta society. Hence, the struggle to maintain their own identity (Balanta means, literally, 'we who know'), implied, especially for the Balanta, the indicated shift to the south and the subsequent decommoditisation. For them it was the only way to reproduce their mode of rice production. And when they succeeded sizeable surpluses were once more achieved. While the northern rice system ceased to produce any rice for the market (in a period in which the productive system as a whole became founded on a 'complete circulation of commodities'), the south instead consolidated its link with the rice market. A considerable part of total production could be turned into a commodity, precisely because the production and distribution of rice were basically founded on non-commodity relations, and geared, above all, towards the reproduction of the productive system itself (and to the reproduction of the involved social relations of production). Hence, it was exactly the relative decommoditisation (as compared to the north) which made it possible in the south to start again and continue the production of rice as a commodity. As a matter of fact, it was the southern rice complex that produced the big boom in rice production and exports in the 1940s and 1950s.

Then a decline followed. We will not discuss this part of the story in any detail here. Suffice it to say that first, the War of Liberation, with its consequent destruction of *bolanhas*, and the increasing labour shortage, then the increasing sahelisation (i.e. the steady decline in rainfall), and finally the almost complete blockage in the supply of consumer goods, all contributed to increasing difficulties in the actual reproduction of the southern rice-producing system.

Production and Reproduction

Figure 6.3 indicates the main flows of rice within and around the household. Although the scheme might appear to be quite complex, it is, in fact, a simplification of what actually occurs.

Rice production and consumption are embedded in a particular chronological organisation. Every cycle of production (and of consumption) crucially depends on preceding ones. This implies, *inter alia*, that a part of current production must be used to offset obligations incurred in previous cycles. In the same way, in and during every (current) cycle of production, the foundations are created for reproduction in the longer term: future cycles are secured. In Figure 6.3 this is illustrated by the amount of rice dedicated to the mobilisation of labour groups, to the organisation of *fanado* ceremonies (in which the younger men obtain the status and knowledge of the elders), and to the accumulation of cattle, etc.

The production and distribution of rice are not only firmly bound together with the particular organisation of time, they are also interlinked with specific forms of the division of labour and, consequently, with specific forms of co-operation. Thus, apparently, it is mainly through rice that households, town and villages, the dry east and the wet west of the country, are bound together. Dry rice (*arroz pam pam*), obtained from Beafada rice growers and used to cover the difficult months immediately preceding the next harvest in the *bolanha* (rice polder), is paid for with 'wet' rice. Rice passes from one family to the other, as compensation for work or

100

Figure 6.3   The main flows of rice within and around the household

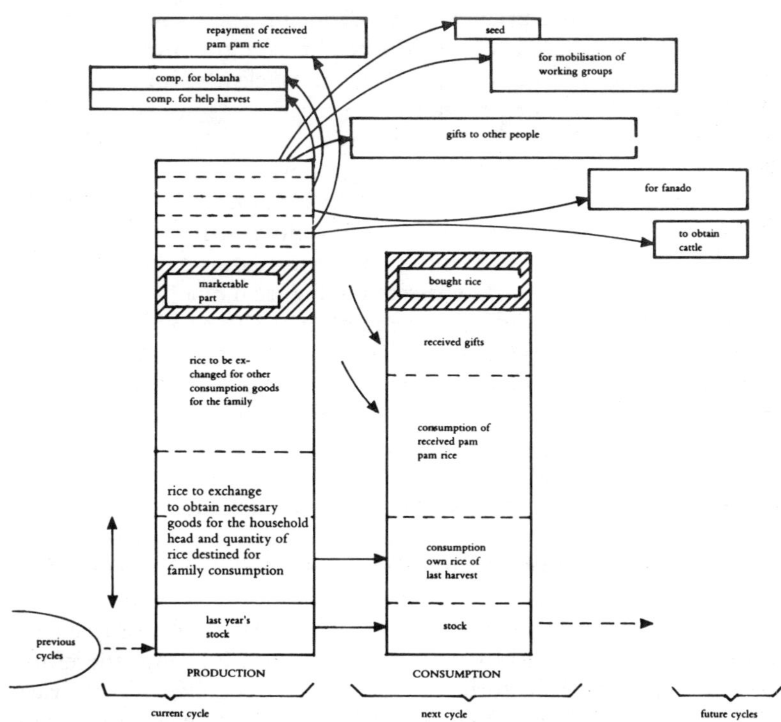

as a 'gift'; rice is handed between men and women, between the
older and the younger generation, between the east and the west: it
goes to the town to 'help' the family there.

Although in most years there is hardly any rice available on the
market, rice is widely distributed. Even such national institutions as
the army jump in with drums full of alcohol to obtain rice.

The individual units of production and consumption are thus
linked together in a complex network of socially regulated inter-
change; simultaneously, this network functions as one of the important social relations of production. To an important degree the
labour process is constituted through it (the mobilisation of labour
groups and their direct or indirect 'payment' with rice is essential to
rice production in the *bolanha*), and the same applies to the distribution of generated wealth (Hochet, 1979; Handem, 1985).

Several additional comments are necessary here. Production and
consumption can never be analysed exhaustively in microeconomic
terms, i.e. by understanding labour, the objects of labour and the

101

instruments involved on the one hand, and the produced output on the other, as commodities, thereby reducing their social value to the (eventual) exchange-value they might have (or might acquire) on the market. In the second place, it is clearly impossible to identify any 'firm' or 'economic enterprise', since the crucial boundary for such an identification is evidently lacking. That is, markets allocating resources according to prices and through monetary exchange and 'firms' within which resources are allocated administratively are absent. At the same time, it must be stressed that the 'market' is not absent – not from history, nor from the contemporary situation. As indicated in Figure 6.3, a (variable) part of production can be sold. At the end of the 1970s and beginning of the 1980s this could rise to 40 per cent of total production; mostly, however, this was much lower, or equal to zero. In fact, the proportion sold varies widely from year to year, from village to village, and from household to household. Interestingly enough, it is mainly the so-called non-commodity factors which promote the selling of rice as a commodity, while an overall commoditisation of the rural economy tends to eliminate rice from the range of commodities. These apparently paradoxical processes are discussed in the next sections.

Simultaneously, it must be stressed that the same complex of relations is characterised by a certain instability. It cannot therefore be seen as a kind of ahistorical model. If, for whatever reason, rice is in short supply for some years, then the very capacity of the system to reproduce itself is reduced. This was evidently the case in the post-liberation war period.

The same happens when, for instance, through specific government interventions, too much marketable surplus is squeezed out; then, too, the reproduction of the whole system and therefore the capacity to produce acceptable harvests in the long run, are jeopardised.

Figure 6.3 demonstrates that rice is crucial for the reproduction of the socio-economic system involved. In the northern rice complex, this role was taken over by money. Consequently, rice production was reduced – the more so since prevailing commodity relations were favourable to groundnuts and relatively unfavourable to rice. But the impact of groundnuts on rice went much further than that.

The Impact of Commodity Relations: Rice and Groundnuts

As previously indicated, the introduction of groundnut cultivation during the colonial period had a marked impact on rice production. Expansion of groundnut cultivation and the concomitant decline of rice production were closely interlinked, especially in the 'northern rice system'. The two not only compete for scarce labour during certain periods, but in addition, through the introduction and growing importance of groundnut cultivation, a new organising principle was also introduced into rice cultivation. That is: *the steady increase of rice production so as to secure the overall reproduction of the particular society (whatever the level concerned) gives way to a more individualised and short-term perspective in which maximisation of cash income per family head dominates.* Such an organising principle will affect rice production, especially when an increasing percentage of labour and land are dedicated to groundnut cultivation.

In 1953, Cabral carried out a thorough agricultural survey (Cabral, 1956). Since the percentage of land sown is specified for each of the ethnic groups, a cross-sectional analysis can be made, which looks at its effects on rice production. Figure 6.4 summarises the results of such an analysis. It shows clearly that the more agriculture is geared towards the cultivation of groundnuts, the lower the yields in rice production. This clearly supports the proposition that the organising principle that dominates groundnut production is indeed extended to rice production as well. Recent studies, based on field research from the late 1970s and 1980s, confirm this interrelation (van der Ploeg and Van Slobbe, 1988).

The Impact of Non-commodity Relations

The same cross-sectional approach can be used to underline the importance of non-commodity relations in the production and distribution of rice. We will do so by presenting some of the results of a number of village studies carried out at the end of the 1970s in the southern regions of Guinea Bissau. Information was gathered concerning village structures (row 1 in Table 6.1), household structures (rows 2 and 3), agricultural activities (in the table symbolised in row 4 as quantity or *busjas* sown per household), etc. The table also contains some statistical elaboration of the data gathered. Running quickly through some of the most important findings, it should be noted that the number of persons per household (per

Figure 6.4 Groundnuts and rice yields per hectare

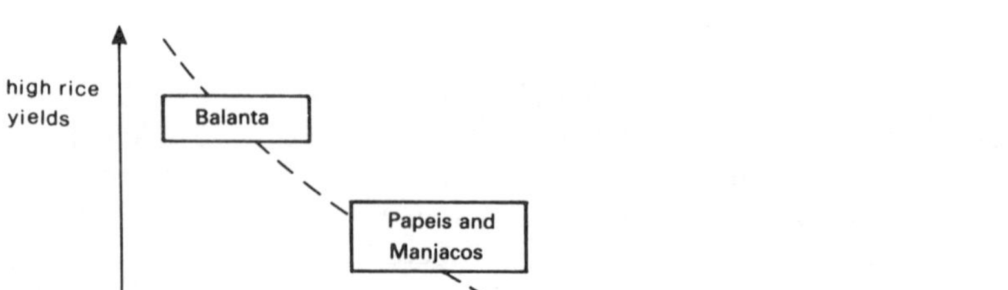

morança) is relatively high ; this corresponds with the findings of Sidersky (1982) and Hochet (1979).

At the same time a considerable variation is to be noted both within and between villages. *Ilheu Colbert* (Balanta) emerges as a notable exception: the average number of persons within the household is less than half that of other villages. As the villagers themselves say: '*i cà tem força*' ('there ain't no power anymore').

To have *força* is essential for the maintenance of the *bolanha* system of rice production. A considerable labour force for the construction and maintenance of new rice polders (or *bolanhas*), and for the preparation of the soil, is crucial for the reproduction of the system both in the short and the long term. So the high average number of persons per household (as well as the high number of households per village) effectively reflects the degree to which 'power' still exists.

The number of persons per household evidently relates to the gender dimension as contained in Balanta society. Women are indispensable for the production of seedlings, for transplanting, harvest and post-harvest activities. Although men may participate in these activities, it is women who control and coordinate these particular labour processes (van der Ploeg and van Slobbe, 1988: 66–9). Balanta women are not bound to their husbands through bride-wealth: they may take off at any time. And so they do. The main mechanism men have to keep wives in their household is to allow them their own fields and, especially, to allow them a sufficient share in the rice harvest. To fulfil all these requirements

Table 6.1 Village and household structures and agricultural activities: Southern regions of Guinea Bissau

	BALANTA villages				NALU village	PAPEL village	BEAFADA village
	Orango	Ilheu Colbert	Gandua	Unal	Calima	Ilheu	Gandua
1. Number of households per village	6	10	16	34	11	25	13
2. Number of people per household	9.3	4.9	11.5	16.0	12.3	8.0	12.2
3. Labour force per household	2.8	1.4	3.7	4.6	2.1	3.4	2.2
4. Number of *busjas* sown	5.5	2.0	6.3	7.8	3.1	3.2	5.5
5. Average *busjas* sown/household member	0.61	0.44	0.85	0.56	0.28	0.39	0.79
(standard deviation)	(0.22)	(0.23)	(1.01)	(0.20)	(0.10)	(0.17)	(1.03)
6. *Busjas*/labour force unit, M	2.08	1.37	2.40	2.02	1.49	1.46	2.99
(S)	(0.85)	(0.40)	(2.15)	(1.10)	(0.71)	(0.60)	(1.89)
7. household members/labour force	3.39	3.77	3.40	3.02	5.78	3.68	4.97
(S)	(1.10)	(1.89)	(1.78)	(1.35)	(4.62)	(1.51)	(1.90)
8. Correlation coefficient: *busjas*, labour force units	0.79	0.81	0.63	0.46	0.40	0.84	0.16
9. Correlation coefficient: household members, *busjas*	0.94	0.76	0.63	0.69	0.84	0.74	0.28
10. Correlation coefficient: household members, labour force units	0.87	0.67	0.89	0.83	0.29	0.81	0.80
11. Correlation coefficient: *busjas*/household members, household members/labour force units	+0.02	−0.69	−0.28	+0.02	−0.51	−0.39	−0.65

considerable rice production is a strategic necessity.

The indicators presented in Table 6.1 reflect the resistance to trends which lead to a decomposition of the social structures that secure the reproduction of the *bolanha* rice production system. A small number of people living in the *morança* points to a high out-migration of young people and women, to a fragmented household structure, characterised by a *homem grande* no longer able to secure food (i.e. rice), status and prospects for the other household members.

A special problem hindering the interpretation of the data in Table 6.1 is the fact that in some villages cultivation in *bolanhas* is giving way to cultivation of rice in the so-called *lugares*. Notable changes in the principles structuring the labour process are associated with this shift, which affect co-operation at village level and, consequently, the distribution of rice among households. There are also some major agronomic changes. As indicated earlier, a *bolanha* is a rice polder on which irrigation is crucial. A *lugar* represents a kind of ecological opposite: it consists of a patch of cleared bush, which is worked individually by one household and is abandoned after some years.

The move to *lugares* is often, though not always, due to socio-economic and/or technical problems in the *bolanhas*. Lack of a labour force or the impossibility of mobilising the available labour power can result in dykes not being maintained. Social conflict within the village might provoke the same effect or prevent adequate preparation of the soil. The same goes for all kinds of technical problems, such as salinisation of acidification of the soil. In these cases, a *lugar* is often sought as a solution. The *lugares* represent an abrupt rupture in the organisation of the labour process. Whereas co-operation at a number of levels dominates in the *bolanhas*, the *lugar* represents an individual organisation of labour, and consequently, an individualised appropriation. So, for whatever the reason (i.e. technical problems or an already disintegrating social structure) the shift from *bolanha* to *lugar* indicates a move from a more collective towards a more individualised means of production. Accompanying this change goes a reduction in technical efficiency. Since no transplanting is done in the *lugares*, and control over all kinds of relevant conditions is reduced when compared with the *bolanhas*, the *lugares* need far more seed to realise a comparable harvest: i.e. the seed: harvest ratio in the *lugares* is lower than in *bolanhas*.

In Table 6.1 two villages represent this shift from *bolanha* to *lugar*. These are *Gandua* (Balanta) and *Gandua* (Beafada): here far more seed is needed to produce a harvest than in other villages.

Row 5 gives the sown *busjas* per household member. This index is one of the best indicators of food security. Assuming there is no drought or severe infestations, the higher this index, the more food will be available per consumer. But again, it should be noted that the two Gandua villages are an exception to this rule. The huge quantities of seed sown do not reflect food security so much as the low technical efficiency of the *lugar* system. As Table 6.1 reveals, the Balanta villages (Ilheu Colbert was excluded for the reason already discussed) have the best ratio in terms of food security. The quantity of *busjas* sown/household member is twice as high or even higher than that of other villages. This underlines the importance of specific socio-economic structures, symbolised here by different ethnic groups.

So far the data have been presented in terms of averages at the village level. Next, analysis can be shifted from the village to the household level. At the household (or *morança*) level a more marked differentiation may become apparent due, among other things, to different families being at different stages of the demographic cycle. Thus, a household with a lot of children but only a few old enough to work, might find it difficult to achieve an adequate ratio of *busjas* per household member. (A *busja* is a container holding 40–43 kg). The last row of Table 6.1 indicates that such inter-household differences exist in non-Balanta villages, as well as in those villages where the *bolanha* is replaced by *lugares*. In the remaining Balanta villages, however, no correlation exists between the labour/ consumer balance, on the one hand, and the index for food security on the other. The potentially disrupting effects of demographic differentiation are compensated for (are neutralised by) levelling mechanisms contained both in the co-operative organisation of the labour process in the *bolanha* as well as in the socially regulated exchange of rice. This is illustrated in Figure 6.5.

The foregoing discussion clearly indicates the crucial importance of non-commodity circuits and mechanisms (such as labour groups, labour exchange and gifts, all following a kind of socially regulated exchange in which long-term reciprocity is central: see Figure 6.3). The more these non-commodity circuits are present and actively reproduced, the more food security is guaranteed, on an inter- as well as an intra-village level. Higher technical efficiency, higher labour productivity and a more equal distribution are secured through the dynamics of non-commodity circuits. Or, otherwise viewed, at least in these typical West African situations, we are led to the conclusion that the more these circuits are broken down, the less food security is produced.

Figure 6.5 An index for food security

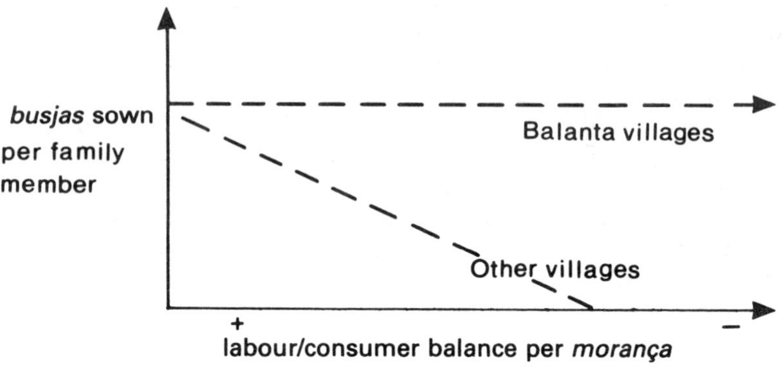

On Technical Innovations

From an agronomic and/or technical point of view, rice cultivation as practised in the *bolanhas* is a highly flexible, dynamic and innovation-generating system. It would be difficult to maintain the interpretation of traditional agriculture as static here. There is considerable opportunity for technical improvements and for endogenous technical innovations. There is much historical evidence in that respect. Cross-section analysis also shows highly flexible and creative adaptation to changing conditions, such as increasing population pressure, land shortages, etc. This can be illustrated in some detail by the techniques for the regulation of water levels within the *bolanhas*.

The base-line solution is quite simple. As demonstrated in Figure 6.6A a simple hole can be dug in the dyke and then, when the tide is low, the excess irrigation water in the field can be drained off. Where labour is scarce, this solution has even been automated, which allows a considerable rise in labour productivity (Figure 6.6B shows that drainage occurs automatically when the level of irrigation water is above a certain point. At the same time the influx of saltwater is prevented, since the construction closes itself when the external water level is high.) A considerable and unsolved problem though in both devices is that they do not allow for much control of internal water levels. As a result, yields within the system cannot be increased beyond certain levels. However, where demographic pressure increases and where sufficient 'força' (i.e. labour mobilisation capacities built on considerable rice stocks) is available, new technical solutions are found, such as the one illustrated in Figure 6.6C. Double dykes make irrigation independent of the external water levels. Both construction and management of this solution require a considerable amount of labour. The major advantage is that considerable rice yields can be obtained, thus creating an improvement in land productivity. The next step (illustrated in Figure 6.6D) is to increase the distance between the two dykes, thereby creating ponds. These can be used both for fish farming as well as for the management of ground water levels and salt balances. This last solution can be found in some Diola villages in the Basse-Cassamance (van der Zaag, 1986).

Ingenious examples of indigenous innovation can also be found in fertilisation, seed selection and the reduction of post-harvest losses. However, what knowledge there is, is not fully exploited. The same is true for innovating capacities: more often than not they are discouraged rather than activated and promoted.

The role of creating, introducing and promoting technical innovation is now largely claimed by state and donor agencies. The current low yields in *bolanha* agriculture are not understood as the result of the partial erosion of the complicated network of social relations of production – an erosion caused *inter alia* by the processes of commoditisation discussed above. Low yields are seen instead as a function of technical backwardness. The internal flexibility as well as the internal room for further technical progress, as demonstrated by the actual diversity which characterised yield levels – the average yield of approximately 1,000 kg per hectare is to be compared with the 3,000–4,000 kg some farmers realise, without any use of so-called 'modern' technologies – are simply ignored by, if not completely unknown to, the planners.

Of course one cannot assume *a priori* that external innovations will not be appropriate to or will not be accepted by the rice growers. History demonstrates the contrary. Those innovations that fit in well with the rationale and subjective logic of the existing farming systems, and which can be acquired on conditions seen as favourable by the rice-growers themselves, will be integrated into rice-growing. An eloquent illustration of the opposite could be found in the growers' attitude during colonial rule: they resisted the use of chemical fertilisers, arguing that they were only meant to squeeze more rice out of them.

The main problem today is that the basic design of the prevailing state- and donor-introduced innovations does not fit in with the specificity of rice cultivation, as illustrated in the previous sections (see, for a more general discussion, van der Ploeg, 1989). This is

Figure 6.6 Techniques for water control in *bolanhas*

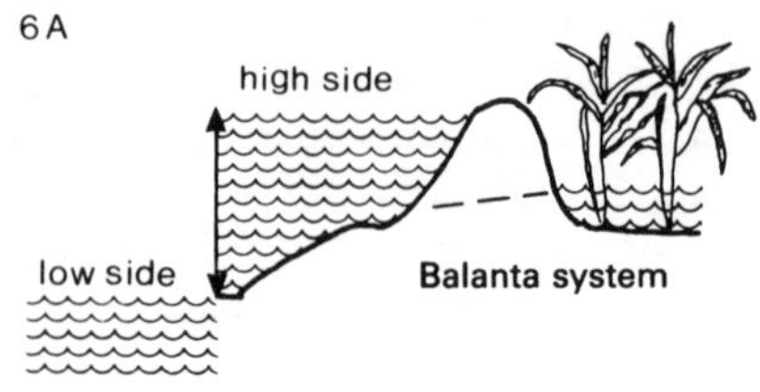

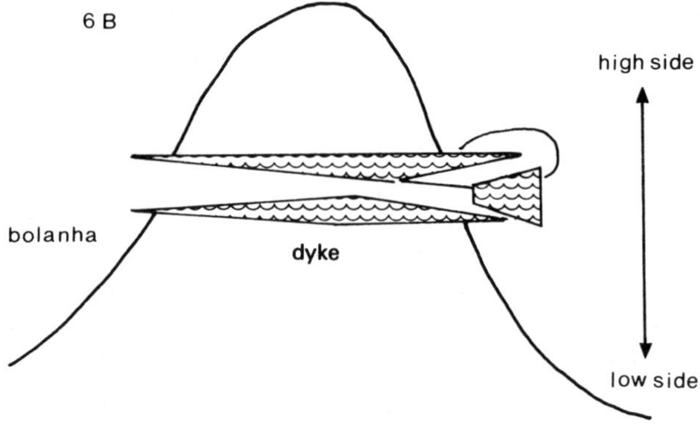

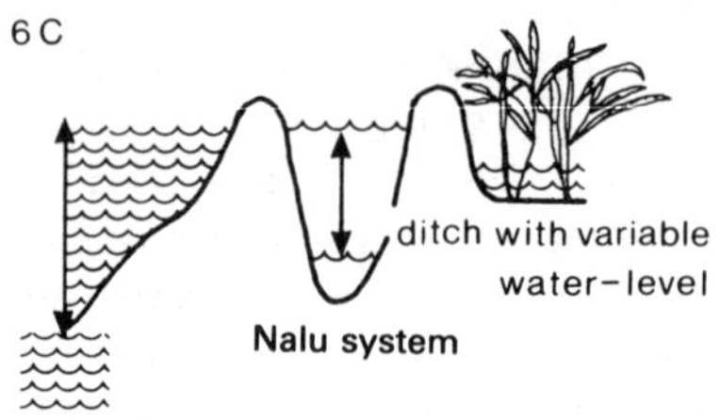

6 D

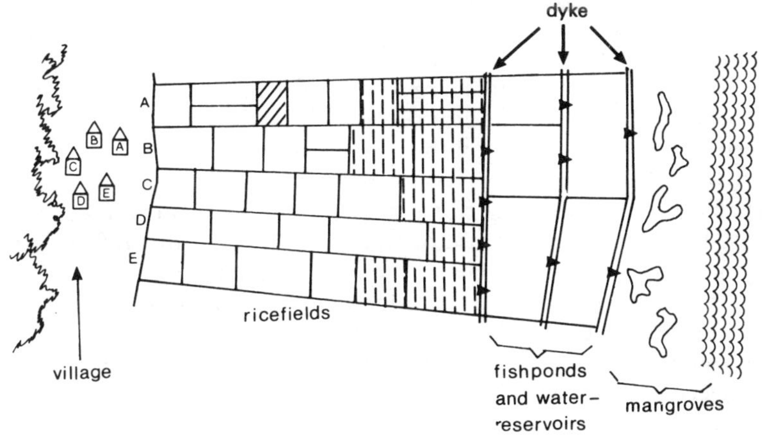

true of both the so-called 'Green Revolution package' and of the now widespread construction of dams. The whole Green Revolution technology is based on the assumption of an immediate and widespread commoditisation of the process of production: of seeds, fertilisers and the means of pest control. 'Increased productivity' is embodied in merchandise that has to be bought. Further commoditisation then often becomes necessary when these new innovations have to be financed with credit. Mechanisation of irrigation is having the same impact.

However, such an outlay for growth and its implied commoditisation runs counter to the specificity already indicated: rice cultivation is based on autonomous, socially regulated reproduction. A reorganisation of this base towards market-dependent reproduction (founded on an expanded circulation of commodities) is either bound to fail, or bound to produce the disappearance of rice as a commodity, as happened to a large degree in the Basse Cassamance (van der Klei, 1989), as well as in the northern rice complex of Guinea Bissau.

The same goes for the central dams, now frequently constructed to improve irrigation and land reclamation in the *bolanhas*. One crucial feature of these very expensive dams does not fit in with local conditions: namely, that the dams entail the centralised management of irrigation and hence of agricultural practice. Until now both irrigation and agriculture have been based on a socially regulated balance between co-operation and autonomy (Ribeiro, 1987). Consequently, reclaimed areas are not used and dams are sabotaged

111

by farmers in all kinds of subtle ways (Hesselink and van Slobbe, 1987). But although a recent French study was published under the telling title 'Barrages contre le développement', dam construction on a large scale continues (Reboul, 1982).

To sum up, one could argue that if a balance between commodity and non-commodity relations is established, i.e. a balance seen as the 'proper one' by the producers themselves, considerable growth rates can be achieved in rice production in the *bolanhas*. A proper understanding of this process, however, is prevented, *inter alia*, by the unreflective use of 'autarky' as a category referring to intrinsic backwardness. Hence the insistence on exogenous technical innovations. But their introduction only exacerbates the 'African drama', especially when, as is often the case, these non-indigenous innovations further disrupt the necessary balance between commodity and non-commodity relations.

References

Baptista, M. M., 'Em prol da agricultura da Guiné, As principais fontes da riqueza agricola. Alguns pontos de critica sobre a sua exploraçao', Boletim Geral das Colonias, Year IX, 1933, nos. 98–9

Bernstein, H., 'Is There a Concept of Petty Commodity Production Generic to Capitalism?' Paper presented at the 13th European Congress for Rural Sociology, Braga, 1985

Bloch, M., 'Economie-nature ou économie-argent: un pseudo-dilemme', *Annales d'Histoire Sociale* I/1, 1939

Boletim da Agencia Geral das Colonias, Guiné, Separata do no. 44, February 1929, Lisbon

Cabral, A. L., 'Recenseamento agricola da Guiné – estimativa em 1953', *Boletim Cultural da Guiné Portuguesa* XI (43), July 1956

Carvalho e Vasconsellos e Monteire Torre, A. B., *Aspectos do problema do arroz na Guiné*, Comissao reguladora de comercio de arroz, Lisbon, 1947

Castro, A., 'Notas sobre algumas variedades de arroz en cultura na Guiné Portuguesa', *Boletim Cultural da Guiné Portuguesa*, no. 19, 1950

Cavallo Viejar, L. A. de, *Guiné Portuguesa*, Vol. II (October 1939), Lisbon, 1939

Ernesto de Vasconcelos, Y. C., *Guiné Portuguesa, Estudo elementar de Geografia fisica, economica e poltica*, Cooperativa Militar, Lisbon, 1917

Graça Correia e Lança, J., *Relatorio da Provincia da Guiné Portuguesa referido ao ano economico de 1888–1889*, Lisbon, 1890

Handem, D. L., *Nature et fonctionnement du pouvoir chez les Balanta Brassa*,

École des Hautes Études en Sciences Sociales, Centre d'Études africaines, Paris, 1985

Hesselink, E. and van Slobbe, L., *Barragems nas bolanhas da Guiné-Bissau*, Sawa, Utrecht, 1987

Hochet, A. M., *Études Socio-économiques conduites dans les régions administratives de Tombali et Quinara; 2me partie: étude socio-économique de la région de Quinará*, Bissau, 1979

Klei, J. M. van der, *Trekarbeid en de roep van het heilige Bos, het gezag van de oudste en moderne veranderingen bij de Diola van Zuid-Senegal*, Amsterdam, 1989

Long, N., 'Creating Space for Change: a perspective on the sociology of development', *Sociologia Ruralis*, XXIV (3/4), 1985

—— and van der Ploeg, J. D., 'New Challenges in the Sociology of Rural Development, a Rejoinder to Peter Vandergeest', *Sociologia Ruralis* XXVIII, 1988

Perspectivas e problemas no desenvolvimento da agricultura, DEPA, Bissau, 1965 (internal document, DEPA, no authors named)

Ploeg, J. D. van der, 'Patterns of Farming Logic: structuration of labour and the impact of externalization. Changing dairy farming in Northern Italy', *Sociologia Ruralis* XXV (1), 1985

—— *La ristrutturazione del lavoro agricolo*, REDA, Rome, 1986a

—— 'The Agricultural Labour Process and Commoditization', in N. Long et al., (eds.), *The Commoditization Debate*, Wageningen Agricultural University, Wageningen, 1986b

—— 'Knowledge Systems, Metaphor and Interface: the case of potatoes in the Peruvian highlands', in N. Long, (ed.), *Encounters at the Interface: a perspective on social discontinuities in rural development*, Wageningen Agricultural University, Wageningen, 1989

—— and van Slobbe, L., *De maatschappelijke organisatie van Waterbeheer en rijstteelt in Guine-Bissau* (The Social Organization of Water-Management and Rice cultivation in Guinea Bissau), Ministerie van Buitenlandse Zaken, DGIS, Den Haag, 1988 (also published in Portuguese by the Ministry of Rural Development and Fisheries, DHAS, Bissau, 1989)

Reboul, C., *Barrages contre le developpement, contribution à l'étude des projets d'aménagement de la vallée du fleuve Sénégal*, INRA, Paris, 1982

Ribeiro, C. R., 'Barragems em bolanhas de agua salgada', in *Soronda 4*, INEP, Bissau, July 1987

—— *Causas de gueda de producâo de arroz na Guine-Bissau*, INEP, Bissau, 1988

Sidersky, P. *Approche de la riziculture Balanta (région de Tombali, Guinea Bissau)*, mémoire DEA, Bissau/Paris, 1982

Zaag, P. van der *Rijstteelt in de Basse Cassamance, Senegal: van surplus tot subsistence*, Wageningen, 1986

7

Staying Together:
Household Responses to Risk
and Market Malfunction
in Mali

Camilla Toulmin

Introduction

This chapter examines the role of risk and market absence or malfunction in explaining household investment in new technologies and patterns of social organisation in a Sahelian farming community in Mali. It starts by looking at the farming system and village community, and examines both the kind of risks faced by people and the reasons why markets for different goods and services do not work well in this region. It goes on to give a brief history of investment by households in wells and ox-drawn plough-teams, shows how production techniques have been adapted to changing market or environmental conditions, and examines why larger households are better able to purchase and maintain these assets. The chapter then examines household size and organisation in these Bambara farming communities, and explores the forces that hold people together within very large domestic groups. Protection from risk and market failure are shown to be major factors explaining such large household size.

The Village of Kala

The area studied lies within the Sahelian zone of Mali, north of the River Niger at Segou (Map 7.1). With an average annual rainfall in recent years of around 400 mm, the village of Kala is close to the northern farming frontier. As is the case throughout the Sahel, rainfall is both highly variable from year to year, and in terms of its

115

Map 7.1 South and Central Mali

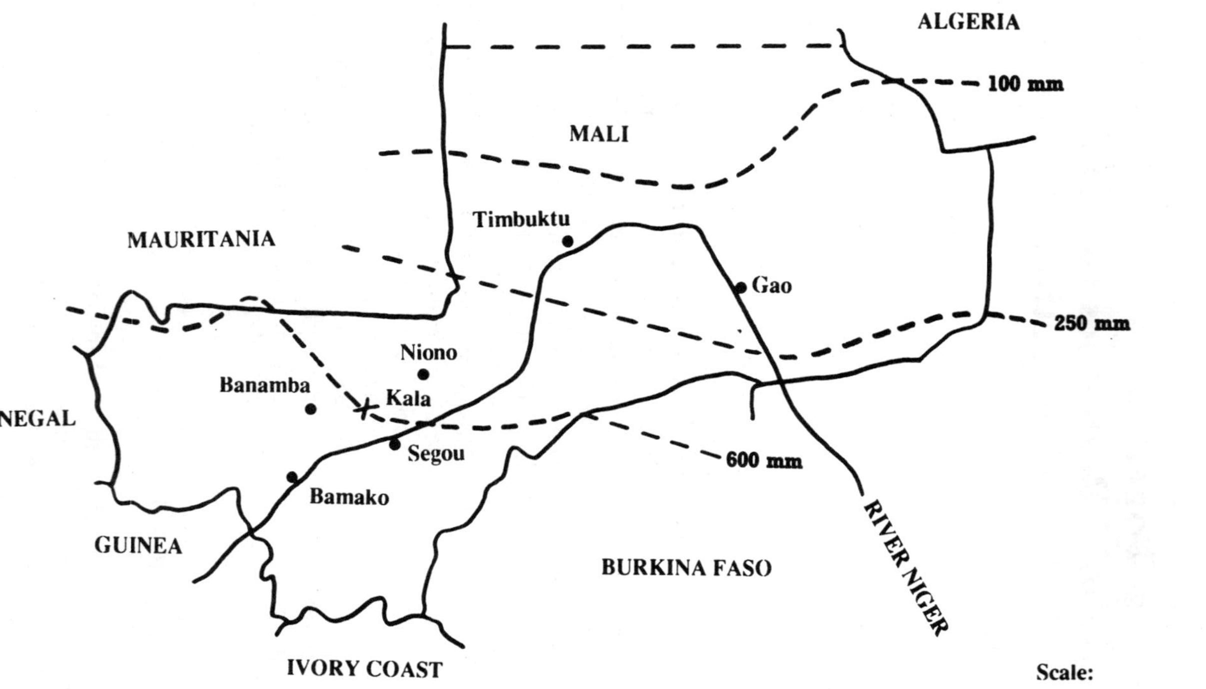

Source: Rainfall data from selected stations, 1920–76.

distribution within the rainy season (IUCN, 1989). Rainfall is also markedly patchy in its spatial distribution, particularly at the beginning and end of the rainy season.

Kala and its neighbouring villages lie on mainly light, sandy soils in a region of relatively low population density. Current levels of occupation are around ten persons/km², although there have been significant levels of migration into the region by farmers in recent years, which is starting to concern the local community. Nevertheless, farmers still reckon that there is plenty of land available for cultivation and that it is labour and plough-team availability that limits their field size, rather than any shortage of cultivable land itself.

The Bambara population of Kala in 1982 was just under 550 people, living in 29 households, or *gwa*. (The *gwa* is the unit that farms a common field and eats most, if not all, of its meals from a common granary.) Table 7.1 shows the distribution of households in terms of size and complexity.

As can be seen from the table, the average size was over 18 people per household, a very large domestic group when compared with village studies elsewhere in West Africa. For example, in Hill's (1972) Hausa study, average household size was 7.2 people, and in Watts' (1983) of northern Nigeria, households averaged 6.9 people. However, other studies among the Bambara yield comparable figures for household size, with Becker (1989) reporting an average of 19.6 people per household for his case-study village to the south-east of Bamako, Mali. Within his sample, a similar breakdown between simple and complex households produces average figures of 10.0 and 31.1 respectively (see also Chapter 8, this volume).

The Farming System

More than 90 per cent of cultivated land is farmed with millet, the remainder being under groundnuts (*Arachis hypogea*), cowpea (*Vigna unguiculata*), fonio (*Eleusine indica*), maize and other minor crops. Two different millet varieties are grown: souna, a short-cycle millet (*Pennisetum* spp, early millet) which takes 60–80 days from sowing to harvest is sown on permanently cultivated, manured land around the village; and sanyo, a longer-cycle millet (*Pennisetum* spp, late millet) which takes 120 days to mature, is grown on large bush fields some distance from the settlement under shifting cultivation; old areas are left uncultivated for long

Table 7.1 Household size by structure Kala, 1982

	No. of households	Average size	Total village population per cent
Complex households	19	23.8	86
Simple households[a]	10	7.6	14
Village total	29	18.2	100

a. Not more than one married couple per household

Table 7.2 Production of two millet varieties, Kala, 1981

	Sanyo bush fields	Souna village fields
Average area per household (hectares)	22.9 (15.0)	5.7 (3.6)
Average yield per hectare (kg of threshed millet)	215 (68.6)	1,010 (395.9)

periods of bush regeneration to restore soil fertility. Two millet varieties are grown in order to reduce the risk of total crop failure, as it is rare for both souna and sanyo to fail in a single year. Risk-reduction by diversifying on- and off-farm activities is a common household strategy throughout this Sahelian zone.

Yields and areas cultivated of the two millet varieties are presented in Table 7.2. Souna produces a much higher average yield per hectare than sanyo, but the area under the former crop is limited by the household's access to manure. Formerly, souna cultivation was limited to very small plots of land just outside the village, but in Kala this crop now covers more than 20 per cent of the farmed area and accounts for half of the total millet harvested.

The shift towards cultivating more souna is due to its short cycle and the greater likelihood that this crop will produce a harvest during the shorter and more unreliable rainy seasons of the past two decades. For souna to produce a good yield, however, it is necessary to manure the land. Manure comes from two main sources: (1) from the considerable numbers of cattle, sheep and goats owned by the villagers of Kala; (2) from the animals of transhumant Fulani and Maure, who bring their herds to Kala during the dry season. These migrants gain access to the water from Kala's many wells by promising in exchange the dung deposited by their animals on the villagers' fields where the herds are penned at night.

The cultivation season starts in June, when the first seed is sown; weeding takes up much of July and August; with the harvest, threshing and storage of both varieties stretching from late September until January. All household members work in the fields from an early age; women start work in the field once the morning's household chores are finished, and bring the midday meal with them. Ox plough-teams are used to ridge land and to weed between the lines of millet. Hand-held hoes are used to clear away remaining weeds.

Groundnut cultivation was previously very important and continued to be so up to the end of the 1960s, after which climatic factors associated with often poor prices combined to reduce greatly the areas cultivated with this crop through much of this region. Groundnuts are now hardly grown at all. While formerly they were of very considerable importance as a source of profit, they often do not now return the seed used, owing to much more erratic rainfall.

There is no road, apart from a dirt cart track, which is used by donkey carts to take grain and goods to and from the market. Steel ploughs from Segou, similarly, were transported on ox-drawn carts the 60 km distance. Ox-drawn carts are now largely displaced by donkey carts.

Risk and Market Failure in Kala

Risk

The main sources of risk facing people in Kala, and the strategies they adopt to reduce their vulnerability, are outlined in Table 7.3. As can be seen, people can only partially protect themselves and their assets from these risks. Large household size and diversified sources of incomes and asset holdings are both central strategies in the face of risk, and are explored later in this chapter.

With respect to human and livestock health, there is little that people can do to protect themselves from loss. The village of Kala is some 12 hours' walk from the nearest adequate health facilities in Segou town. In rare instances, people seek such medical attention, but it is effectively unavailable for most people. The costs of

Table 7.3 Risks and household strategies in Kala

Risks	Strategies
Rainfall variability and crop failure	Grow two crop varieties
	Diversify off-farm incomes
	Migrate in years of failure
	(to relatives, town, etc.)
Loss of assets	
Cattle: theft	Employ herder
disease	Little veterinary medicine available
over-milking	Visit and monitor herder
Wells : collapse	Regular maintenance
Ploughs and carts	Regular maintenance, careful use
Human illness and death	Little modern medical care
Infertility	Traditional medicine
	Buffer of large group

treatment are high in terms of medicines, transport, and food and lodging for other family members accompanying the sick person. Access to veterinary care is even more limited, and is restricted to periodic campaigns against rinderpest.

The isolation of Kala and neighbouring villages from almost all government services is also exemplified by the absence of any formal educational provision in the village. In 1988, however, two men from Kala were given training in basic literacy and they now act as teachers to the rest of the community.

Market Failure

Many of the goods and services needed by people in Kala are not available on the market. In addition, there are no insurance markets that can help provide a cushion in the event of loss. Market absence or failure can be accounted for by various factors. First, Kala lies in a region where land is in abundant supply, with labour being the scarce factor which determines how much land can be cultivated. There is little labour available for hire. In the short cultivation season, people prefer to work their own land, rather than hire themselves out to work for others. Considerable effort is made to reserve whatever food is available for the farming season, so that households do not have to send labour out to earn extra food. Only in a few cases were households forced to sell some days of labour to another household in order to supplement grain stocks. Overall, less than 10 per cent of days spent farming in Kala were provided by hired labour. The absence of a labour market means that households are very largely dependent on their own labour for farming. The long-term strategy of household heads is thus to ensure that resources are put aside to meet costs of marriage and that members of the household feel happy about remaining part of the group. Fostering of a boy or girl is a possible medium-term solution to lack of labour, if a couple lacks children or when children are still too young to work in the field, but sooner or later the fostered child will return to his or her natal household. Fostered children cannot be formally adopted by another household.

Second, the market for other farm inputs is limited by the desire of households, where possible, to acquire control over these assets themselves. The two most important inputs into farming in Kala are dung and plough-teams. Dung is obtained only by providing well water to a livestock keeper. No cases were known of either dung or water being sold for cash, though elsewhere in the Sahel such transactions do take place..Dung is as valued by farmers as is a good dry season water supply by herd owners, and consequently a direct exchange has traditionally been established between these commodities. It would be practically impossible for a farmer to pay a herder to manure his fields without offering him water in exchange, since the herd could not stay around Kala without access to water. Thus, if a farmer wants access to more dung, he must get another well dug.

In the case of ox plough-teams, there is a limited degree of hiring which takes place, as shown in Table 7.9 (p. 130). However, in general, people much prefer to own or control a plough-team themselves. This is because the value to the farmer of the plough-team is much greater in the few days immediately following a heavy rain, a time when it will be impossible to borrow a team from another household as the latter will be busy at work in their field. Farmers dependent on hiring equipment have to wait until plough-team owners have a spare moment to help them plough and weed their land. Richer households did not have a surplus of animals and equipment acquired with a view to hiring out; well-equipped households used them for expanding their own field size. There seem to be problems with ox hire markets in many other contexts, because of the risks to oxen from poor handling, and the high cost to the owner of effectively monitoring the proper care of his animals. When oxen were hired to another family, these were usually elderly animals of relatively low value, lent prior to their sale for meat.

Insurance markets for certain services, such as cover against crop failure, are frequently lacking both in the developing and the developed world. There are certain inherent difficulties relating to the economics of information and moral hazard which make crop insurance difficult to monitor and to assess how far the particular event (e.g. crop failure) is the result of risk (e.g. rainfall failure) or due to the actions of the person insured. Nevertheless, some levels of cover are provided to farmers in the developed world; this, however, does not exist in most parts of Africa. Similarly, life assurance and pension services are not available to most people in Mali, due in part to low levels of service sector development and low levels of income. People must provide their own insurance against the risk of their own or others' death by reducing the impact of death. They do this by giving each other mutual aid and support within the context of the large Bambara *gwa* or household. Members of the *gwa* have certain obligations to each other, but they also have rights which ensure them a certain level of care and support when they are in need.

The market in medical care, as noted above, is limited by the perceived ineffectiveness of traditional medicine for many ailments and the difficulties of access and cost for modern medicine.

High risk and market absence or failure thus provide the back-cloth against which people have to choose which crops to grow, how to plan their investments and how to arrange their family life.

Investment in Kala: Wells and Ox Ploughs

Farmers in Kala have a range of investments to make. Some assets improve the productivity of their farming system, such as wells and ox plough-teams; some provide longer-term security to the individual and household, such as stores of millet and holdings of cattle, sheep and goats; others enable a diversification of incomes, such as investment in a trading business or in a donkey cart, used to transport goods and passengers. Individuals and households must also invest in raising the long-term viability of their domestic group, through payment of bride-wealth and marriage costs.

Wells and ox plough-teams are the two major farm investments made in Kala. Well-digging is a traditional skill throughout this region; and recent years have seen a great expansion in the number of wells being dug in many villages like Kala, in order to benefit from water–dung exchanges. Ox plough-teams were first introduced into the Segou region in the 1930s and have gradually

122

become widespread. Other assets, such as bicycles and carts, were first acquired by people in this region in the 1950s, while investment in a small shop in Kala has only occurred since the end of the 1970s.

Wells

Wells have always played a role of great importance in this large, arid zone north of the River Niger, as can be seen in the abundance of oral legends relating to the discovery of drying-up of water sources. Until Independence in 1960, only the village chief had the right to dig wells within the village territory, so households other than the chief's did not have their own private wells. Independence brought a change in the law whereby all Malians were entitled freely to exploit land and water resources, and as a result, well-digging by individual households in Kala gradually became established. Many village households are now independently involved in manure–water exchange contracts with visiting pastoralists during the dry season.

The first well belonging to a household other than the chief's was dug in 1963, followed by several others over the next few years. By 1979, more than half of the village households (15 out of 29) had their own private wells and by 1982, 75 per cent of households (22 out of 29) had their own wells. The dry season of 1983 saw a further large increase in the number of wells dug. In addition to private wells, there have always been a number of public wells, of which as of 1988 there were still five used for domestic water supplies. One public well, 20 metres in depth – the *kolonba*, or big well – to which the founding lineages have priority access for their herds, is deep enough to water several cattle herds throughout the dry season.

Table 7.4 shows the distribution of wells among households in Kala. It will be seen that by 1982, only 7 out of the 29 households in the village were still without a well of their own, these seven being the smallest and poorest households in Kala. Key characteristics of these households are presented in Table 7.6 and discussed below. Table 7.5 shows that in 1980–1 the number of private wells grew by 69 per cent, from 16 to 27. If this growth in wells dug seemed impressive enough during the period of research, the subsequent period has seen an even greater increase, with a total of 16 wells being dug in the single dry season of 1983 – an increase of 60 per cent – from 27 to 43 private wells. The community did not consider there to be a significant lowering of the water table as a result of the

Table 7.4 Well ownership in Kala, end of 1981

	No well	One well	Two wells	Total
No. of households with well	7	17	5	29
No. of private wells	—	17	10	27
Of which:				
dug in 1980		4	2	6 (22% of total)
dug in 1981		3	2	5 (19% of total)

great number of wells recently dug. Wells tended to be spaced fairly widely apart; but there is a progressive decline in the water table as the dry season lengthens, with a rapid and substantial recharge at the start of the rainy season. Thus, for example, water is only 4–5 metres from the surface during the rains. By the end of the dry season, it is commonly more than 30, sometimes 40 metres from the surface in some of the deeper wells. Many of the herders draw water using an ox to pull a water skin that takes some 20 litres or so. Others have a team composed of three young men to draw up the bucket. Watering a herd of 60–70 animals usually took 1–2 hours, depending on the season.

Four of the 11 wells dug in 1980 and 1981 were second wells for households wishing to expand the area cultivated with village field millet. Most of these households have their own cattle herd which takes much of the water from the existing well, thereby limiting the quantity available for establishing manure–water contracts with visiting pastoralists.

Table 7.5 presents data on the number of wells dug in the two dry seasons of 1980 and 1981, the status of households digging these wells and the means with which the wells were dug. Larger households used their own labour for well-digging, while smaller households had to hire labour for this task. Of the 16 wells dug in the dry season of 1983, only 2 were first wells, 10 were second wells and 4 represented a third well for certain households.

Differences in the speed with which households have invested money and labour in digging a well are a result of:

1 *The cost*: well-digging requires an outlay of capital to finance the purchase of pick-axes, hire of a blacksmith to dig the well if household labour is not to be used and to provide for the work party which will reinforce the well-head. These initial capital costs averaged between 60,000 and 80,000 Malian francs (1,000

Table 7.5 Well-digging in 1980 and 1981

No. of wells	11	
No. of Households	11	
for whom first well	7	
for whom second well	4	
Undertaken by:		
household labour	7	(mean no. of workers 10.5)
hired labour	4	(mean no. of workers 4.1)
paid for: by livestock sales	2	
by grain sales	2	

MF = 500 FCFA = 10 French francs); and annual maintenance costs of 12,000 MF must also be paid. For the smallest, poorest households, the cost of well-digging has been insuperable. They lack the labour force to dig the well, and they also lack those assets that could finance the hire of a blacksmith to do the work.

2 *Livestock holdings and date of settlement*: households differ in the date of their settlement in Kala. The earliest settlers (17 out of 29 households) have prior access for their herds at the one public well large enough to water stock. Households whose predecessors arrived later in the village, during the nineteenth century, must wait for the founders' cattle to be watered before their own herds can drink. For a later arrival with a large cattle holding there are strong incentives to get a well dug early, as the herd can then be watered at dawn and spend the day at pasture.

3 *Household size and political strength*: in the early years following Independence, a household digging a private well faced considerable hostility from the chiefly lineage, despite the change in the chief's legal position in the matter of control of water resources. The first well to be dug after 1960 was by the leading rival to the chief, belonging to one of the other two founding families. This man was able to provide a test case in getting a private well dug, not only because of the family's long settlement in the village but also because his household was sufficiently large and wealthy to withstand intimidation. The main households to follow this man's lead in the subsequent period of well-digging also belonged to the largest and strongest families in the village and could also claim to have been among the first settlers of the village.

Returns from Well-digging   The main returns to a household from digging a private well come from increased yields of the short-cycle millet, souna, as a result of the household gaining more access to dung, through water-manure exchanges. An analysis of the net returns from well investment for households of different size has been presented elsewhere (Toulmin, 1986). This analysis showed the large benefits to be gained from digging a well and the short payback period for this investment. It is no surprise, then, to witness the eagerness with which all households in Kala have tried to get a well dug, even at considerable personal cost.

Consequences of not Owning a Well   Table 7.6 presents the characteristics of those households in Kala still without their own well in 1982. The figures reflect both the causes and consequences for households without wells of having little dung to boost yields of the short-cycle millet.

When considering whether to dig a well households must take into account not only the relative returns to wells as opposed to other assets, but also the longer-term consequences for the household of persistently low grain production if it fails to get a well dug.

Ox Plough-teams

Ploughs were first introduced into this region in the 1930s with the establishment of the Office du Niger cotton-growing irrigation scheme around Segou and Niono. Two households from Kala bought ploughs in the early 1950s, in the form of the heavy steel equipment sold in Segou. More widespread adoption followed the fabrication by local blacksmiths of a lighter ridging plough, which is all that is required for sandy soils and light enough to be pulled by local oxen. Ploughs were first bought to prepare land for sowing groundnuts and village field millet. They could ridge land much faster than the hoe so that a larger area could be sown within the limited time available. Weeding ploughs are a more recent development and are used to break up the soil between the lines of millet, speeding up the work of weeding the extensive area farmed with this crop. Draught oxen work in pairs. However, many households have three oxen per plough and rotate the animals in turn. With a three-ox team they tend to work during the peak season for three or four hours, in the morning and evening. If they have fewer oxen they work a somewhat shorter time. Thus a plough-team day is between five and eight hours long. Millet was not fed to draught oxen or other cattle.

Distribution of Plough-team Holdings   Table 7.7 refers to those

Table 7.6 Characteristics of households without wells

	Average households without walls	Average village households
Household size:		
No. of people	6.9	18.2
No. of workers	3.5	7.5
Cattle per household	1.7	20.8
Total millet production		
per worker 1981: souna	538 kg	807 kg
sanyo	478 kg	634 kg
Total per worker	1,016 kg	1,541 kg
Total per person	515 kg	635 kg

Table 7.7 Plough-team holdings Kala, 1981

	At least two oxen and plough	one ox and plough	one plough	Nothing	All households
No. of households	26	1	1	1	29
Per cent of households	91	3	3	3	100

households with permanent access to plough-team equipment and animals in the farming season of 1981 and thus includes those households which had borrowed oxen for the entire rainy season. As can be seen in Table 7.7, only one household had neither a plough nor draught oxen. A second household had invested in a plough, which could be lent to another household, loan of the plough being repaid by the loan of oxen. The third household without a complete plough-team had bought a plough in early 1981 and was loaned a young ox for its use in return for help in watering stock during the dry season. The majority of households had at least two oxen and a plough at their disposal during the farming season of 1981, and many had several teams of oxen.

Most households in Kala have one or two plough-teams, but the three largest households operated four or five plough-teams at the height of the weeding season. As may be seen from Table 7.8, the number of plough-teams operated is strongly related to the size of the workforce.

Purchase of an ox plough-team requires an investment of at least

Table 7.8 Distribution of households by number of work-oxen owned in 1980 and 1981

| | No. of Oxen | | | |
	0	1–4	5–8	9–12
1980	3	19	4	3
1981	3	18	5	3
Mean household size (no. of workers 1981)	2.6	5.3	13.0	16.3

Note: In 1980, of 109 oxen used, 8 (7.3 per cent) were on loan from other households. In 1981, of 120 oxen used, 11 (9.2 per cent) were on loan from other households.

140,000 MF in the case of a pair of 2 to 3-year-old oxen, which are bought for training and use the following year. Purchase of a full-grown and trained pair doubles the initial investment cost.

Maintenance costs for the oxen and equipment are also high: oxen must be watered throughout the dry season, herding fees paid and the plough blade renewed. The opportunity cost of providing labour to water animals for the long dry season is the greatest maintenance cost. Where the household has a large cattle herd, the unit watering cost per ox will be low, but for a household with a single pair of oxen, the cost in forgone migration earnings of keeping a young man at home to water cattle over several months was about 25,000 MF.

Alternatives to Purchase of a Plough-team Households rarely buy a complete ox plough-team in a single year because the overall expense is very great. Instead, the team is acquired over a period of several years and in the interim animals are borrowed from relatives and neighbours.

Loan of oxen: in 1981, 11 out of 120 work-oxen in use in Kala were on loan from another household, for the entire cultivation season or for a large part of it. Young oxen are often lent for a couple of seasons in exchange for their being trained by the receiving household. In other cases, an ox may be loaned to a household in exchange for labour to help with watering stock during the dry season. Households that rely on borrowed oxen are in an insecure position as the owner may demand the animals' return at short notice or may decide to sell them. In addition, the loaned animals are rarely highly productive, being either young and poorly trained, or beasts at the end of their working life.

Hire of plough-teams: there were few cases of plough-team hire in Kala in 1980–1, reflecting the fact that all but three households had permanent access to a team in both cultivation seasons. Details of how these three households obtained the use of a plough-team are presented in Table 7.9, from which may be seen the variety of sources tapped and terms obtained, with much of the help freely given. In straight exchange, the hire of an ox plough-team for one day is usually paid for by two or three days spent weeding with a hand-hoe (equivalent in wages to 6–8 measures of grain, worth 1,500–2,250 MF: the unit weight of a measure is around 1.5 kg of millet).

Two principal reasons account for the low level of plough-team hire in Kala and the emphasis given to having permanent access to a plough-team during the farming season:

1 The short rainy season and the erratic distribution of rainfall mean that the usefulness of a plough-team varies considerably from one day to the next, being of highest value in the few days following a heavy rainfall when land can rapidly be prepared and sown. A household that needs to hire a plough-team often will not get it at the optimal time, because the oxen are busy at work in the field of the owner, and it may be worth little to get the team at any other time.

2 The low level of plough-team hire is also due to the fact that bush field millet, sanyo, is usually sown on unridged land, so it is not essential for a household to gain access to a plough-team in order to cultivate this crop. Nevertheless households without permanent access to a plough-team did suffer lower yields of bush-field millet because of the long time taken to weed those fields by hand.

Consequences of not Investing in a Plough-team   When deciding whether to invest in their first ox plough-team a household in Kala must consider not only the flow of costs and benefits from the whole investment, but also the speed at which the investment should be made (given the possibility of borrowing a spare ox from another household and the possibilities of hiring a plough-team from others if no part of the investment is made). Table 7.10 presents data on yields of millet for households without permanent access to a plough-team. Their low village field yields (souna) are mainly due to their also lacking a well and therefore having no dung for fertilising these fields. Low yields per hectare of the bush field millet are due to slow and incomplete weeding of these fields.

Table 7.9 Access to a plough-team for poor households, 1980 and 1981

1980

Household *x* received	8.5	*plough-team days*
of which:	3.0	days freely given by neighbours
	4.0	days in exchange for 8 days' weeding
	1.5	days in exchange for 2 man- and 4 woman-days' weeding
Household *y* received	6.0	*plough-team days* plus a youth to guide the oxen in exchange for the loan of *y*'s ridging plough and 6 man-days of weeding
Household *z* received	6.0	*plough-team days*
of which:	5.0	days in exchange for 12 days spent by *z*'s son guiding the oxen for the plough-owner in the latter's field
	1.0	day freely given by a Fulani

1981

Household *x* received	12.5	*plough-team days*
of which:	5.0	days in exchange for a month's help in watering cattle by *x*'s son
	4.5	days in exchange for 8 days' weeding
	3.0	days in exchange for 3 days' work by *x*'s son and daughter weeding the plough-owner's field with the plough team
Household *y* received	2.0	*plough-team days* freely given by friends, relatives and neighbours

Household *z* having bought a plough in the dry season of 1981 was lent a young bull to be trained by a relative and was able to borrow a second ox for a total of 26 days, all of which were freely given.

Note: Plough-teams are of special value in preparing land for village field millet and where teams have been hired, it is this crop which takes priority.

The Pattern of Asset-holding and Investment by Households

Investment in a well and plough-team is currently of central importance to the household's investment strategy since these assets allow the household to maintain and raise their millet yields. Only

Table 7.10 Yields of millet for three households without their own ox plough-teams

	1981	
	Sanyo bush field millet (kg/ha)	Souna village field millet (kg/ha)
Mean for village	215	1,010
Household x	97	1,088
Household y	262	–
Household z	144	665

Note: Household y had no well and a very poorly manured village field, and therefore chose to concentrate its small workforce on the bush field millet, sanyo.

once these assets have been acquired do households turn to investing surplus in breeding stock. However, the pattern of asset-holdings must be seen not only in terms of the current returns and risks pertaining to different assets, but also against the background of changing economic and technical conditions. Historically, cattle were the main form in which surplus could be held and wealth in cattle was limited to a few households in the village. The introduction of groundnut cultivation in the 1940s provided a means to generate a surplus, enabling many to finance the purchase of ox plough-teams and to build up cattle herds. Short-cycle crops have now become crucial in supplying grain for household needs and a surplus for investment in livestock. High yields are dependent on a regular supply of dung. Wells have therefore become the latest productive asset in which households want to invest, since these allow them to maintain and expand the area they farm with short-cycle millet.

Table 7.11 presents the distribution of households in terms of their asset-holdings and demonstrates a strong positive correlation between ownership of different assets. At the top left-hand end of the table are found households without wells and few cattle apart from a pair of work oxen, while at the bottom right-hand end, all households have at least one well and they keep large holdings of both work oxen and breeding cattle. The strong association between livestock holdings and household size can also be seen from the mean size of household for each column and row of Table 7.11.

Table 7.11 Distribution of wells, work oxen and breeding cattle holdings, Kala, 1981

Holdings of work oxen	Breeding cattle holding				
	0	1–10	11–20	21–40	40
0	2ᵃ 1ᵇ (2.6)				
1–4	4ᵃ 5ᵇ (4.9)	1ᵃ 5ᵇ 1ᶜ (5.1)		1ᵇ 1ᶜ (7.9)	
5–8			2ᵇ (10.7)	1ᵇ (11.6)	1ᶜ (14.9)
9–12					1ᵇ 1ᶜ (16.1)
13–16					1ᵇ 1ᶜ (16.9)

Note: Figures in the matrix refer to the number of households.
Figures in brackets are the mean number of household workers for that element in the matrix.
a. Households with no well.
b. Households with one well.
c. Households with two wells.

Changes in Asset Ownership over Time

The position of a household at any one time in terms of the composition of its assets represents a point reached within a dynamic process, the result of past performance in agriculture and of decisions to convert assets from one form to another. Over time, a household will hope to move towards the bottom right-hand end of Table 7.11, investment in productive assets (work oxen and wells) providing a surplus to invest in breeding cattle. Household size plays a crucial role in this process of accumulation. Larger units are better able to generate the investment sum required for well-digging and the purchase of ploughs and oxen, thereby overcoming any problems posed by indivisibility. Larger households can also diversify their production activities and take risks with new technologies early in the 'adoption cycle'. A large labour force, containing men, women and children of various ages, is itself an asset of value and part of a household's income and capital are devoted to expanding the size of the workforce of the domestic group.

Households at the top left-hand end of the table are handicapped not only by their lack of assets, but also by their small size. For them, diagonal movement in the table is constrained by their inability to generate sufficient surplus to invest in raising farm productivity. With their chronic grain deficits, the cohesion of the domestic unit is threatened each year once the household granary is empty and individuals are left to their own resources. Digging a well has been essential for enabling some of these households to produce enough millet for their domestic needs, as they lack their own cattle to provide them with dung. Households with a plough-team but no well have generally not been able to produce a regular supply of millet, depending as they must on lower and more variable yields of longer-cycle millet for the bulk of their needs. Some of these households have seen a progressive worsening of their fortunes, shown by their movement towards the left-hand end of the table, as cattle have been sold to pay for grain.

Bambara Household Organisation

The Bambara household was described earlier as that unit which farms a common field and eats from the same granary. The large average size of these units was shown in Table 7.1. Here we examine why people live in these large domestic groups and assess the overall strength of forces making people stay together. It is shown that there are substantial advantages reaped by larger households in the fields of production and investment. Establishing and maintaining a large domestic group with a diverse portfolio of assets are also important ways to cope with the many risks to which people are subject and the problems of securing access through markets to many goods and services. The potential drawbacks to individuals from belonging to a large group, such as loss of control over one's labour and resources, are minimised by defining the specific duties of each individual to the household and benefits to be received in exchange. Flexibility within these contracts between a household and its members allows the household to respond to changes in internal and external conditions that threaten its unity.

The size and structure of the household at any one point in time is the product of several factors. Fertility and mortality rates are important determining variables, partly under the control and manipulation of household members. Household size is also the product of the relative strengths of opposing forces leading either to the formation and cohesion of a large, extended group or to its division into several separate units.

Advantages to Large Household Size:

(i) Areas of Co-operation

There are several benefits from the organisation of labour and resources within a large household in communities like Kala. Larger households are better able to generate and maintain a surplus and to ensure their reproduction in the longer term for the following reasons.

Diversification of Income Sources This reduces risk since variability in different incomes is not likely to be perfectly correlated. In Kala, some diversification takes place in the farming sector, with the cultivation of two millet varieties of different cycle length. However, diversification outside the farm sector is of equal or greater importance. A profitable off-farm income and household wealth tend to be mutually reinforcing, surplus generated in one area being used to raise productivity and expand production in other sectors. The following are examples of diversification:

1 In the rainy season, women in larger households have time to plant a small plot of maize and millet; this supplements household foodstocks and provides a private source of income for the woman. In none of the five smallest households did any woman of working age have a private grain field of her own, whereas all women in other households had their own private fields.

2 In the last few years, several households have detached one member to run a full-time trading business during the farming season, a period when village traders have a near monopoly on the supply of goods in the village. Out of the six permanent trading businesses, five are in the largest and richest households, for whom it represents an additional source of income and avenue for accumulation. The sixth case is a man of seventy from a poor household who, while trading full-time, has very low stocks and deals in low quality fish and peppers, rather than the more profitable kola-nuts, kerosene, sugar, etc.

3 In four of the largest households, an older man was regularly sent out hunting for game, contributing to the household's productivity by giving them better food rather than remaining with them in the field.

Opportunities for income diversification in the dry season last from the end of the grain threshing in January until the first sowing of crops in June. Advantages from large household size arise during the dry season because the labour force can be divided among various activities, some of which contribute to the household's farm income while others represent avenues for income earning. Examples of these are:

1 Women share the household tasks where they are not the sole woman in charge of cooking, leaving time free to devote to spinning cotton bought at Segou market, collecting bush produce, dressing other women's hair, petty trade, etc.

2 In the larger households, several young men can migrate to town, leaving a few of their peers behind to water cattle and prepare land for the new farming season. Migrants' earnings help reduce pressure for sales of grain or livestock to meet tax, marriage and other expenses.

Vulnerability to Demographic Risk Larger households are more robust in the face of demographic risk in the following ways:

1 Vulnerability of farm production to the illness or death of a working member will be greater where there are fewer workers.
2 A balanced sex ratio will be more likely where there are several couples producing children.
3 A domestic group containing couples of different ages will exhibit a less marked cycle in its ratio of consumers to workers.
4 A larger household is more likely to contain a greater spread of labour categories, each with their part to play in household production and organisation, from the five-year-old given the infants to guard, to the retired man or woman left to care for the sheep and goats while the rest of the family is at work in the fields.

Economies of Scale in Agriculture

It might be expected that there would be some economies of scale in farm production, given the number of duties and operations which have to be carried out during the short rainy season. The data for Kala show no overall tendency for yields of grain per worker to increase with increasing household size. While yields are lowest in the smallest households, this is due to their lower holdings of assets, such as wells and plough-teams, rather than to diseconomies of farm production. To the extent that economies of scale in farm production exist, they probably are obtained at a fairly low level, in moving from a nuclear household to one containing two or more men and their wives. Having access to different kinds of labour seems more important than size of household alone. Nevertheless, larger households can accumulate an absolutely larger surplus, which can finance investments such as oxen, ploughs and wells.

Acquisition and Maintenance of Farm Assets

Household size is relevant to two assets in particular – wells and cattle. The only households left in 1982 without a well were the smallest, poorest ones. Households with sufficient labour have dug their own well while smaller households with other assets have sold these to pay for a well to be dug. In the case of both plough oxen and breeding cattle, at least one young man must be kept at home in the dry season to join the water-drawing team. In smaller households, there is often only one young male worker, and his obligation to stay at home during the dry season imposes a heavy cost on the household, since it limits his freedom to earn cash as a migrant worker.

Examples of the advantages of large household size include:

1   In both 1980 and 1981, 15 households out of the 29 in the village ran out of millet early, before the October harvest of souna was finished. The average size of these millet-deficit households was 12.7 people as compared with the village mean of 18.2.
2   This greater ability to ensure sufficient food supplies for the household in the larger domestic groups is reflected in the greater ability of individuals, particularly women, to establish private holdings of sheep and goats. In poor households, these animals must frequently be sold to buy grain when the household granary is empty. The four households with the largest numbers of sheep and goats contained an average of 41.8 people and had an average of 2.15 animals per person. This contrasts with the mean household size for the village of 18.2 persons and a holding of 1.27 animals per person.

(ii) Areas of Conflict

As households grow in size, certain problems arise due to labour management, monitoring of others' effort, distributional issues and the reduced family feeling that exists between more distantly related kin.

The labour incentive problem is met in Kala by defining a certain sphere within which individuals owe all their time to the joint enterprise of the domestic group, the product from which provides them with their basic food needs. The power to exact this labour effort is backed up by incentive and force. Incentives include the approbation of others in being hard-working, while disincentives include the loss of access to food from those not part of the work group, as food is only prepared and served to this group. Young men who do not work face ridicule, and jeopardise the return obligation of the household to provide them with a wife.

Distributional issues arise over either individual wealth being used for communal purposes or the converse, the appropriation of household goods for an individual's benefit. For example, livestock owned by an individual are not available to finance general household needs. In one case, an old woman's ox had been sold and the proceeds used to buy grain for the household. This was acknowledged to have been wrong and the old woman was provided with another animal.

A further force pulling people together in large groups is the extreme weakness of a single person if having to cope on his/her own in this society. Both men and women face a very insecure and difficult future if they separate themselves from the wider household. A man depends on his father to arrange and partially pay for the costs of his marriage; he does not himself have access to marriageable women, at least not those from traditional sources. As a single man, he would find it almost impossible to survive: cooked foods are not available to buy in villages like Kala, unlike in much of Hausaland where numerous meals and snacks are prepared for sale. Even when he marries, the small nuclear family unit he may head is highly vulnerable in the event of the illness or death of one of its members: it may also not rear successfully the children needed to maintain the group in the future. Labour and other farm inputs are not easily available for hire, as explained earlier, making such a nuclear unit highly dependent on its own stock of assets. It is unlikely that it would be able to create the surplus to invest in such assets or have the labour to maintain these successfully.

Women's fallback position is even weaker than that of men. They rarely have any power to determine where or with whom they will live. Once married with children, a woman will usually stay with her husband, whatever the circumstances, at least until her children are old enough to be independent of her. However, by this stage she will increasingly need their help and the support with which they provide her.

Household division among these Bambara farmers is neither a regular nor a predictable stage which a man can expect to experience at a certain point in his lifetime, such as on the death of his father. Instead, many households in Kala have remained together over several generations, despite the deaths of succeeding household heads. In Kala over the last thirty years, there have been only five cases of division out of the current twenty-nine households. In none of these cases was division prompted by the death of the household head. Details of these cases are shown in Table 7.12.

That household division is a relatively rare event is due to several

Table 7.12 Household division in Kala over last thirty years

Households	Date	Circumstances
1A/1C	early 1960s	Separation of two full brothers from main household. Said to have been personality differences but 1C is also a viable unit of several men and women.
1A/21	early 1960s	Illegitimate man forced to establish himself as a separate unit on his marriage.
2A/2B	late 1960s	Separation of three full brothers from main household, both units containing 15–20 people at the time of division, though 2B has subsequently declined in size and fortunes.
4A/4B	mid-1950s	Separation of two full brothers from larger group. Said to have been obstinate and wanted independence.
15/16A	mid-1970s	Separation of man from half-brother's son; former probably engineered split to enable him to take large share of household cattle herd.

factors, some of which have been mentioned earlier. First, individuals perceive there to be strong advantages to remaining part of a large domestic group. Second, division of a large household can be postponed by renegotiating the terms of the implied contract binding people together. Third, there is strong ideological value placed on remaining within the single household; feelings of shame are associated with the division of households, since it implies that the interests of these individuals could not be accommodated within the large group. Fourth, while the overall population growth rate for the region of which Kala is a part is currently estimated at around 2.5 per cent per year (Randall, 1984), many households have not seen their own numbers grow at that rate, so the anticipated regular division of households due to their growth in size has not taken place.

Conclusions

This chapter has described the choices faced by Bambara farmers in the fields of investment and domestic organisation. These choices about production, new technologies and investment in a range of assets are made within a context of high risk and inadequate markets. People have adapted strategies for coping with this situation, which involve diversifying incomes and assets. In the domestic sphere, diversification takes the form of creating and maintaining large domestic groups which buffer the individual from demographic misfortune.

References

Becker, L. C., 'Conflict and Complementarity in Bamana Farming: a case study of Soro, Mali', PhD thesis, SOAS, London, July 1989

Hill, P., *Rural Hausa: a village and a setting*, Cambridge, 1972

IUCN, *Sahel Studies*, International Union for the Conservation of Nature, Gland, Switzerland, 1989

Randall, S., 'The Demography of Three Sahelian Populations: Marriage and childcare as intermediate determinants of fertility and mortality', PhD thesis, London University, 1984

Toulmin, C., 'Changing Patterns of Investment in a Sahelian Community, DPhil thesis, Oxford University, 1986

Watts, M., *Silent Violence: food, famine and peasantry in northern Nigeria*, University of California Press, 1983.

8

Population and Change in a Gambian Rural Community, 1947–1987

Margaret Haswell

This chapter reviews from a baseline of empirical evidence of pre-existing roles the new roles that households assume in response to technological and environmental changes over time. Successive population censuses for the years 1949, 1962, 1973, 1982, and 1987 indicate that agricultural methods and property relationships are subject to continual change, and identify reallocations in household human resource use when adapting to successive change – new combinations of family enterprise in a constantly changing social and political environment, complex problems of household inequalities in capabilities and opportunity, as a new generation reaches maturity with a different set of goals in its sights.

In 1947, in the then relatively isolated Middle River area of The Gambia, the people of Genieri village practised shifting cultivation of plateau lands supplemented by rice grown on annually flooded river flats. More favoured sandy loams fringing the river were no longer cropped under shifting cultivation but were cultivated continuously. In this initial period, however, the Colonial Medical Research Council's choice of Genieri as a suitable location for investigation of nutritional deficiencies in Africans to be financed from Colonial Development and Welfare Research funds, disrupted traditional lifestyles.

The Village, 1947–50

Gamble (1955) describes the customary rights of the village head in a typical Mandinka village as follows:

> When a village is established the head of the founding lineage chooses
> land for farming and housing for himself and the heads of families who
> settle with him also select tracts of land with his knowledge and consent.
> Thereafter, their descendants have the right of usage, the right to lend
> the land, and in some cases the right to give it away. . . . Should a
> lineage move or die out, land it used can be reallocated by the village
> head either to those who are short of land or to new settlers. But once
> land has been granted to a Compound Head, the Village Head has not
> the right to take it back.[1]

Traditional agricultural practice involved a number of operations
which required different intensities of energy output or work.
Work on crop production was sharply divided between men and
women. The men concentrated their energies mainly on growing
groundnuts for sale, and were responsible for the production of
millet foodgrain working together in comparatively large fields.
All who ate out of the same pot grew food together. Groundnuts,
on the other hand, were grown by individuals on small plots of
which they owned the usufruct. The choice open to the women
responsible for production of upland rainfed and swamp rice, and
Digitaria exilis harvested during the 'hunger-gap' between old and
new crops, was determined mainly by the size of the female labour
force. In smaller households women tended to concentrate on
upland rainfed rice to avoid the long and tedious walk to the
swamp. However, age and sex compositions did not fully explain
wide differences in the choice of crops between households. There
were also differences in attitudes towards women's work compared
to men's work, to cash compared to food crops, to early crops
compared to late, to risky crops compared to surer crops, and in
sickness of individual members of the household at critical times.

'Scarcity of season', however, was an overriding factor in inter-
preting use of labour and returns to labour. The rainy season is
short, lasting from June to October followed by a long dry season.
Labour in the busy seasons soon after the rains break, and during
the first weedings in particular had opportunities of being much
more productive than labour at other seasons.

By custom, and as a result of the spread of Islamic culture, the
people of Genieri are a polygamous society, and in 1947 the
extended family system was strongly entrenched. The farming
economy was organised within the framework of the compound,

---

1. D. P. Gamble, *Economic Conditions in Two Mandinka Villages*, Colonial Office,
1955.

family households normally including one or two married brothers or married sons of the compound head.

Mechanisation Experiments and their Effect on the Pattern of Labour Use

On 18 May 1949 the Colonial Governor, in a despatch to the Secretary of State, considered the prospects for the employment of machinery in the agriculture of the territory. These he regarded as:

> depending on the results of the experiments of the Field Working Party at Genieri. The Party's programme is to make a thorough study of one small area over a period of years in order to determine the choice of a balanced diet which a village community can produce for food, and to experiment in the mechanised production of the necessary agricultural crops. First, the operations are conducted not on 'plantation' principles, or for the achievement of a model show-place, but as a matter of partnership between the technical officers in charge and the villagers of Genieri. Secondly, the main criterion to be applied to the experiments is whether any pattern of mechanised cultivation can be developed which village communities could operate from their own resources as a paying concern.[2]

Map 8.1 shows the site of the Nutrition Field Working Party's camp in relation to Genieri village, and delineates areas where mechanisation trials with tractors and standard farm equipment were conducted as an aid to land and crop improvement; it was assumed that tractor work for preparation and planting would give more time for hand weeding, and lengthen the period over which weeding could be done effectively. It had already been shown that earlier planting combined with good weeding and a higher plant population gave better results.

Since more land was available than could be managed with the hand-hoe alone, the villagers identified suitable areas for the experiments which would *not* disrupt the traditional farming system, namely:

1 Plot M, covered by a deciduous seasonal forest in various stages of regeneration and unsuitable for the mechanisation of

---

2. In reply to the Secretary of State's Despatches No. 25 of 22 February 1947, and No. 121 of 13 July 1948.

Map 8.1   Genieri village lands: 1948–9 mechanisation experiments

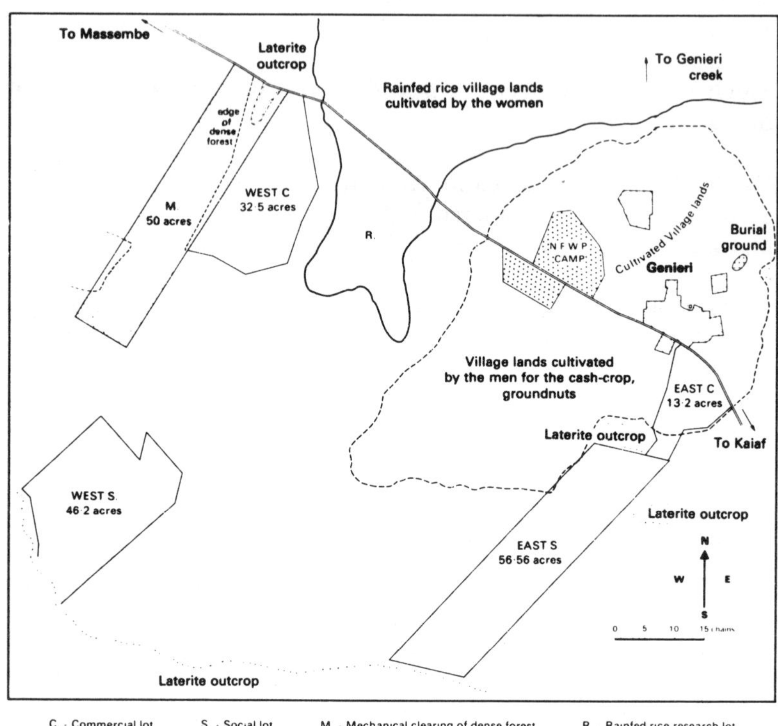

Source: Prepared by Cox Cartographic Limited, 1987, from map drawn to scale by the Nutrition Field Working Party in 1948.

Note: 1 acre = 0.4047 hectares.

permanent field crops, for experiments in low-cost bush-clearing operations.

2   Plots C, where laterite outcrops were found to underlie the surface by no more than 5–15 cm, and where the irregularity of the slope made anti-soil erosion measures – an integral part of the mechanisation trials – difficult, for experiments in strip-cropping on the contour with a wide variety of food crops.

In 1948 seventy-nine Genieri farmers pooled an area of 'old tillage' (Plots S), hired tractor and other services from the Field Working Party, and planted the cash-crop groundnuts; but the return from the sale of groundnuts was meagre compared with the almost insuperable problems of recruiting a large force of voluntary labour, and the experiment was not repeated in 1949. By this date, however, the mechanisation trials were providing year-round employment for some members of family households who were paid in food and cash.

Land and Labour Use for Crops under the Indigenous System

In 1947 a detailed study of farm family household activities had identified what constituted 'good' and 'bad' farming within existing bush–clearing and hand–hoe practice;[3] in 1948, in the prevailing climatic regime of short wet and long dry seasons, a marked loss of bodyweight of both men and women was recorded as foodstocks ran out and the period of heaviest agricultural work began in preparation for the new season's crops. About 30 per cent of the villagers showed signs of hunger or famine oedema at this time.[4]

In that women and girls were responsible for rice production, and rice, by weight, made up about 80 per cent of the total foodgrain supply, the female labour force of individual households determined in large measure the choice of crops and food supply. Men in households with a relatively large female labour force tended to devote less time than other households to clearing a field for millet in guinea savannah above the laterite ridge (Map 8.1), and more time to producing groundnuts for sale. Groundnuts for cash, and swamp rice for consumption within the village, were the first and second choice of crop.

The average yield of groundnuts for the whole village in 1949 was 541 kg per hectare (0.83 kg per hour of labour). With the existing pattern of cultural practices, without material change in these practices, no very much greater intensity of labour used was likely to have been profitable in that year. In fact, the average practice secured a reward of 0.91 kg of groundnuts per hour of *additional* (or *marginal*) labour, and growers were not, *on the whole*, prepared to use more labour if they received for it returns of less than 0.91 kg of groundnuts per hour. The average return was probably also the marginal return per hour for late millet grown

3. M. R. Haswell, *Economics of Agriculture in a Savannah Village*, HMSO, 1953.
4. Colonial Medical Research Committee Third Annual Report, Colonial Research 1947–8, Colonial Office: HMSO, Cmd. 7493, p. 68.

Table 8.1 Average returns to labour in Genieri, 1949

	kg per hour
Late millet	0.47 grain
Early millet	0.49 grain
Sorghum	1.07 grain
Digitaria exilis	0.99 grain
Maize	0.91 grain
Rice-upland rainfed	0.17 paddy
Rice-swamp flood-irrigated	0.64 paddy
Groundnuts	0.83 nuts in shell

above the laterite ridge, since land was not 'scarce'; but in that land for early millet, sorghum, and maize (grown within the fenced area of compounds only) was in short supply, the average returns for these crops were probably very much higher than the marginal returns. Since groundnuts could be grown comparatively successfully on guinea savannah soils of varying quality which were not regarded as 'scarce', costs of labour and returns to labour were more important than yields per hectare of land.[5]

The period for which the vote for the Nutrition Field Working Party at Genieri was approved ended in the spring of 1950, and responsibility for the station was transferred to the Colonial Government.

The mechanisation experiments in 1948 and 1949 had shown that the use of the tractor was extremely costly and uneconomic; and that its introduction in savannah country 190 km from the coast, where there were no roads and no tractor agents or service facilities, had required the provision of a workshop (sited in the neighbouring village of Massembe; see Map 8.1) on a scale out of all proportion to the scale of the project. Attempts to introduce anti-erosion measures and strip-cropping had resulted in disappointing yields after the almost insurmountable problems met with in rearranging 'pocket-handkerchief' field and plot boundaries to which particular heads of households laid claim. The presence in the village of Europeans who provided alternative employment opportunities for younger men and boys resulted in many households neglecting their food-crops; and communal institutions were threatened, notably the age–sex organisations, the main function of which was the supply of labour to overcome 'bottlenecks' in some farming operations.

5. Haswell, *Economics of Agriculture*, pp. 56–7. See also Appendix Table 8.7.

'It is of considerable interest', the Colonial Medical Research Committee reported after the departure of the Nutrition Field Working Party, 'that during 1950 the mean weights of the men and women of Genieri village have fallen from the highest values found in the three years of their activities to the same lowest value recorded in the autumn of 1947.' It had been foreseen that should land supplied for mechanisation experiments be returned to Genieri farmers, modern methods of production with their high capitalisation and potentially higher returns would stop, and local methods, with their negligible capitalisation and known economic returns, would take over.[6]

The Study-area Resurveyed, 1962

The Colonial Office report on The Gambia for the years 1950 and 1951 records that 'the progress of development has been greatly assisted by the establishment in 1951 of the Farmers' Fund from the profits of the newly constituted Gambia Oilseeds Marketing Board'.[7] By the mid-1950s twenty-nine projects were receiving grants from the Farmers' Fund, mainly on schemes related to rice development, one of which was awarded to Genieri for an access causeway to enable Genieri women rice farmers to reach more fertile swamps near the mangrove edge. These tidal swamps are salt during several months of the year and cannot be double-cropped.

One of the subjects of enquiry in our resurvey of Genieri was the quantity of grain and groundnuts available per head in the community for the year 1961–2. The estimated mean values of the ratio: total quantity of grain and groundnuts produced by the community over total population of the community measured in man-equivalents, indicated that total product from agriculture was approximately 30 per cent higher in 1961–2 than in 1949–50. Significantly, this was largely contributed by the smaller compounds, because many of the women in these female labour-scarce households who had earlier avoided the long and tedious walk to the swamp now used the causeway to reach more fertile soils from which they were obtaining very high yields of rice; and it will be seen that the rate of growth of *food* production was just about keeping pace with the rate of growth of population.

However, pressure upon lineage heads, caused by the increasing

6. Ibid.: 137–41.
7. Ibid.: 49.

Table 8.2 Growth rates per cent year, Genieri, 1949–62

Rate of growth of population:	
Male	2.2
Female	1.0
Total	1.6
Rate of growth of number of households	5.2
Rate of growth of agricultural	
production per head of population:	
Food crops	2.2
Total agricultural production	2.9

Source: Haswell (1975: 133).

tendency for fission from the extended family to form nuclear family units, had reached a point of almost total abandonment of guinea savannah areas previously bush-cleared and fired by the men of the household for late millet. Half the population now lived in households containing fewer than twenty persons compared with 25 per cent in 1949; and only in the three largest households (which averaged forty-seven extended family members) did the men working together clear an area for late millet in 1961 – a total of 4.9 ha compared with 35.8 in 1949 (Appendix: Table 8.7). This change in the pattern of crop production had three important effects:

1 The non-availability of male labour to clear a field for late millet, valued more highly than rice in the period of heaviest agricultural work, and the almost total dependence on rice as the staple foodgrain, subjected the community to possible deficiency of B complex vitamins. To make the porridge easier to swallow and more palatable, millet-eating communities in West Africa also usually supplement the cereal with relishes – meat, fish, caterpillars, locusts, ants, and both wild and cultivated vegetables, to which a groundnut sauce may be added.
2 The greatly increased area of flood-irrigated rice following the construction of a 4 km causeway to more fertile soils nearer the mangrove-lined bank of the river, saw a change in work-sharing roles, with men now helping the women during transplanting, the chief bottleneck.
3 Women had surplus rice to sell, and purchased clothing, which had traditionally been the responsibility of the men, and which the men began to neglect; but if the supply of foodgrain for

home consumption fell short before the new harvest they were held responsible for replenishing stocks.

At this date, work was in progress with the construction of an all-weather road from the capital Banjul on the coast to Mansa Konko 190 km up-river on the south bank connecting with the trans-Gambia highway, which would eventually pass close to Genieri (Map 8.2, p. 160); and the rate of population growth in Genieri was accounted for by recent immigrants (Appendix: Table 8.5). They included Arabic teachers, masons and petty traders; and the indigenous community gave new settlers the usufruct of land on which to grow groundnuts.

Genieri village had been designated one of the Gambia Oilseeds Marketing Board's sixty Official Buying Stations, although the greater proportion of the groundnuts purchased were grown by farmers in neighbouring villages. The Board distributed seednuts at planting time in June, and received in return five bushels for every four bushels issued as soon as the trading season opened. The Board's Licensed Buying Agent, resident in Genieri for part of the year, engaged in retail trade for the United Africa Company, and provided credit at times of seasonal food shortage when the farmers pledged their groundnut crop. Loan repayments were made when the groundnut crop was sold in the following January. There was also extensive borrowing from traditional village moneylenders and petty traders when interest rates averaged 73 per cent per annum. The producer price of groundnuts each year was therefore of primary concern to the farmers. At this date, the price of groundnuts relative to the price of rice was about half, an economic equivalent well below the calorific equivalent in which values are broadly comparable.

In 1961 almost all the smaller households had borrowed for foodgrain. Cereal shortage had become acute towards the end of September, and continued until the main rice harvest at the end of December. The 'famine' food *Digitaria exilis*, harvested in September, supplied grain for only three weeks; and the inadequacy and poor distribution of rainfall for the early rainfed rice crop, planted on fringing clay soils at the margin of the swamp, resulted in negligible yields for all households.

A few years earlier, in 1955, Gambia Department of Agriculture had launched their 'oxenisation programme'. A farmer using ox-drawn equipment could cultivate four or five times the area of a hand-cultivated farm, but the main contribution of animal power was to the cash-crop, groundnuts, grown on guinea savannah (sandy) soils by men of all ages, because seednuts could be obtained through the marketing board if they had none or had insufficient from the previous harvest. However, Genieri people were not a cattle-raising community, and in 1961 only one farmer from a small household used animal power. He hired an ox plough-team from neighbouring Kaiaf village (Map 8.1). None the less, a few better-off households had begun to invest in cattle, purchased from the *Fula* (or *Turanko*), a millet- and cotton-growing, cattle-raising people living on Gambia's southern border with Senegal; and these they tethered to manure their groundnut plots about four to six weeks prior to planting. The highest yield obtained in 1961 on manured plots was 1,772 kg/ha nuts in shell, more than double highest yields obtained on unmanured plots in 1949.[8]

Second Resurvey of the Study-area, 1973

Hand-hoe agriculture had now given way to ox-ploughing over large tracts of guinea savannah, and the decrease in area under cultivation on these soils in 1962 over the 1949 figure was reversed with continued neglect of foodgrains (millet and sorghum) in favour of groundnuts grown for cash; but further causeways had been constructed giving access to more fertile mangrove swamps with increased areas under rice – still heavily dependent on the female labour force. During 1973, however, widespread drought in many areas of neighbouring Sahel countries had drawn the attention of international bodies, and a massive relief operation was launched.

In Genieri, the area under upland rainfed rice had been reduced, and salt encroachments in the rice-growing swamps were being recognised as a more serious problem. Only 20 per cent of households had sufficient stocks of grain from the previous year's harvest to tide them over until the first harvest of short-term crops, *Digitaria exilis* and rainfed rice; and in August the village made a bulk purchase of imported rice through their area council for resale to individual households. Borrowing for food became a high proportion of household debt. By this date, 32 per cent of households contained fewer than ten family members. Only 16 per cent of households contained more than twenty people, namely founding lineage households in which the 'wealth' of the village was

8. M. R. Haswell, *The Changing Pattern of Economic Activity in a Gambia Village*, HMSO, 1963, pp. 43–55. See also Toulmin in this volume.

concentrated, owning 73 per cent of the capital value of all live-stock, including breeding cattle, pairs of draught oxen and horses. These were households (including moneylending members) that had invested in improved living quarters with corrugated-iron sheets replacing thatch roofs, households that had demonstrated an ability to turn inherited rights over land to advantage, and households which had retained command over human resources and acquired a range of local skills. Twenty-four per cent of the population, averaging 8.4 persons per household, still possessed no livestock and lived in traditional mud and thatch huts.

Only three households possessed a horse, and these also had private wells. The horse not only competes with man for food-grain, but also for fresh water (about 45 litres/day). By comparison, the water requirements of a draught ox averages about 15 litres/day – rather more during periods of work, rather less when not working and left to forage in the bush during the long dry season. Recommended regular feeding of grain, a ration of 3 kg millet on days when working (equivalent to about 0.5 kg/day/year, the subsistence requirement of one adult person), was rarely adhered to; and acute seasonal shortage of grazing after the end of the rains had resulted in groundnut haulms, previously discarded at the time of threshing, acquiring a market value for livestock feed.[9]

Almost all use made of animal-drawn equipment was applied to the groundnut crop on guinea savannah (sandy) soils. Lawrence and Smith (1988) state that:

> oxen on poor quality diets cannot increase their consumption of food to match the energy they spend while working. . . . This might be the underlying reason for the failure of single-ox cultivation in Ethiopia. Where the quality of food is very poor, it is often better to have two oxen doing what little they can without losing too much weight rather than to have one ox which soon becomes exhausted beyond recovery.[10]

Lawrence points out that there is little correlation between power output and energy used with reference to the high energy cost of movement: even when ploughing nearly 40 per cent of the total energy for work is used just for walking. 'What does seem to be fairly consistent', he states,

> is the amount of energy that oxen are willing to expend each day and this

9. M. R. Haswell, *Energy for Subsistence*, London, 1985, p. 58.
10. P. Lawrence and A. Smith, 'A Better Beast of Burden', *New Scientist*, 21 April 1988, pp. 81–2.

depends mainly on body weight, the level of feeding, and management. So long as the draught force exerted by the animal is sufficient to move the required implement faster than about 0.5 m/sec. then everything should be O.K.; but if the implement is so heavy that the draught force exerted by the animal is sufficient only to move it at speeds of less than 0.5 m/sec. then either a different implement should be used, or more animals. Very few oxen will work well at such low speeds and most just give up altogether if asked to do so.[11]

Genieri farmers used draught oxen in pairs. Three of the 'rich' households owned two pairs and were able to put the second pair to work as the first pair became exhausted, thereby lengthening the working day when timeliness of the operation was critical. Many of the poorer households without draught oxen now owned a donkey, the traditional pack animal in The Gambia. Said by Genieri farmers to be the cheapest draught animal to maintain, donkeys do not roam the bush but are kept within the household compound and consume domestic waste discarded by the women when preparing meals. The water requirement of the donkey is also lower, about 5 or 6 litres a day. Again, the weight of the implement is important; donkeys pull the lighter groundnut planter (Appendix: Table 8.11). Donkeys and horses were, however, mainly used for carting threshed groundnuts from the field to the buying station in Genieri village. Clark and Haswell (1970) give a range of transport costs in developing countries, expressed as kilograms of grain equivalent per ton km transported, and arrive at a median for each of several methods used: porterage 9.0, pack animals 4.6, wagons 3.4, boats 0.9 and motor vehicles 1.0.[12] Transporting harvested paddy rice from the mangrove swamp along the causeways continued to be by porterage.

Little advance from the subsistence agriculture production of the 1940s had been made in more than two decades. Yields of swamp rice were no longer significantly higher with further causeway expansion into more fertile mangrove areas. Almost all use made of animal-drawn equipment was applied only to the cash-crop groundnuts grown on Guinea Savannah (sandy) soils. Early millet, which requires much labour bird-scaring, had been virtually abandoned, and late millet and sorghum were no longer grown in pure

11. Lawrence Peter, Royal (Dick) School of Veterinary Studies, Centre for Tropical Veterinary Medicine, University of Edinburgh, private communication, 25 May 1988.
12. C. Clark and M. Haswell, *The Economics of Subsistence Agriculture*, London, 1970, p. 203.

Table 8.3   Total production contributed by individual crops

	1949–50 %	1961–2 %	1973–4 %
Food-crops			
Late millet	2.0	0.6	2.6
Early millet	0.4	—	neg.
Sorghum	0.2	—	0.2
Digitaria exilis	1.6	1.4	1.0
Upland rice	6.4	5.8	4.8
Swamp rice	57.9	54.9	45.4
Cash-crop			
Groundnuts	31.5	37.3	46.0

Source: Haswell (1975: 165).

stand, except by the one or two remaining large households which still had sufficient adult males to clear a foodfarm. Instead, single rows of millet and sorghum, providing a meagre return, were interplanted in the groundnut fields of some households (Table 8.3 and Appendix: Table 8.11).

The low yield–base of late millet (*Pennisetum typhoides* or pearl millet, commonly referred to as bulrush millet) grown under African conditions was receiving major attention at the International Crops Research Institute for the Semi-Arid Tropics. Cummings quotes an average yield for Africa in 1972 of 733 kg/ha, compared with 1,735 kg/ha for Latin America, 945 kg/ha in the Near East and 460 kg/ha in the Far East, whereas yields of 3–4 tons/ha are obtained in North America and Oceania. Among the major objectives of the Institute, which was chartered only in July 1972, is that it should serve as a world centre to improve the genetic potential for grain-yield and nutritional quality of sorghum and pearl millet. 'Millet', Cummings states:

> can be ratooned to produce a sequence of fodder crops or of fodder crops followed by a grain crop. In this respect it is superior to sorghum. . . . The utilisation of dwarfing genes to pull down the plant height has the potential of making it possible to pack larger numbers of plants on to a given area of soil and to increase the ratio of grain to stover and thus improve the efficiency of conversion of photo-synthetic energy into immediately usable food products.[13]

13. Cummings was Director of the International Crops Research Institute for the

153

Cummings had expressed concern about the real problems which arise for the farmer when a succession of good years is followed by a period of long-continued drought (as was experienced in the 1970s): 'no significant breakthroughs in production technology for the semi-arid tropical agricultural areas comparable to those for wheat, rice and other cereals in the areas with assured water supply had occurred.'

The 1949 survey of traditional practice had shown that density of plant population coupled with cleanness of weeding were factors determining yields of late millet. Little-weeded crop yielded only 114 kg grain/ha compared to 407 kg on much-weeded areas with a high millet plant population.[14] The relatively high ratio of energy gained to energy expended on hand-cultivated early millet (Appendix: Table 8.8) was more a consequence of low labour inputs – mainly bird-scaring by children – than of high yields, which were no better than for late millet in the absence of controlled use of dung (this volume, Chapter 7); but this calculation did not include energy expended in hand-pounding harvested grain. Eastman (1980) observes that traditional methods for making flour are extremely energy demanding. A woman can devote 2–5 hours a day processing sorghum and millet grain for her family – time and energy which could be more profitably spent on other activities.[15] Neither did the higher rice yields, which followed with deeper penetration into mangrove swamps after the construction of access causeways, materially improve the ratio of energy gained to energy expended, because of the high overall human energy costs borne by the women who were attempting to produce and process virtually all the food requirements of the household unaided.

The very low growth rate per cent per year of the indigenous population (Appendix: Table 8.5) is determined by the shape of the survival curve (Figure 8.1). The wet season not only coincides with maximum disease transmission and marginal food supplies, but also with a sharp reduction in the time given to child-care through intense agricultural activity, which is reflected in high infant mortality.

By 1973 the village was within reach of a surfaced road and had access to groundnut marketing through co-operatives. The co-operative had for some time permitted the issue of loans at a low fixed rate of interest to groundnut producers (the men of the

---

Semi-Arid Tropics, Hyderabad, India.
  14. Haswell, *Economics of Agriculture*, pp. 52–3.
  15. P. Eastman, *An End to Pounding*, Ottawa, 1980.

154

Figure 8.1   Survival rates in Genieri, derived from age distribution of deaths, January 1962–June 1973

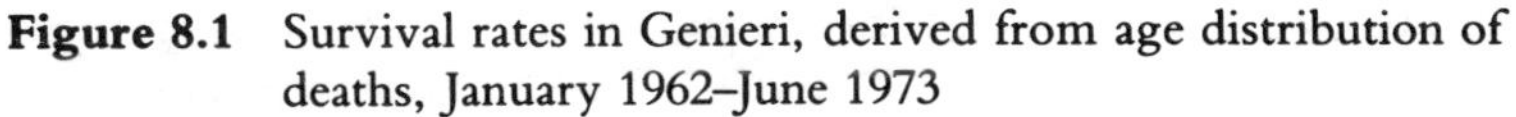

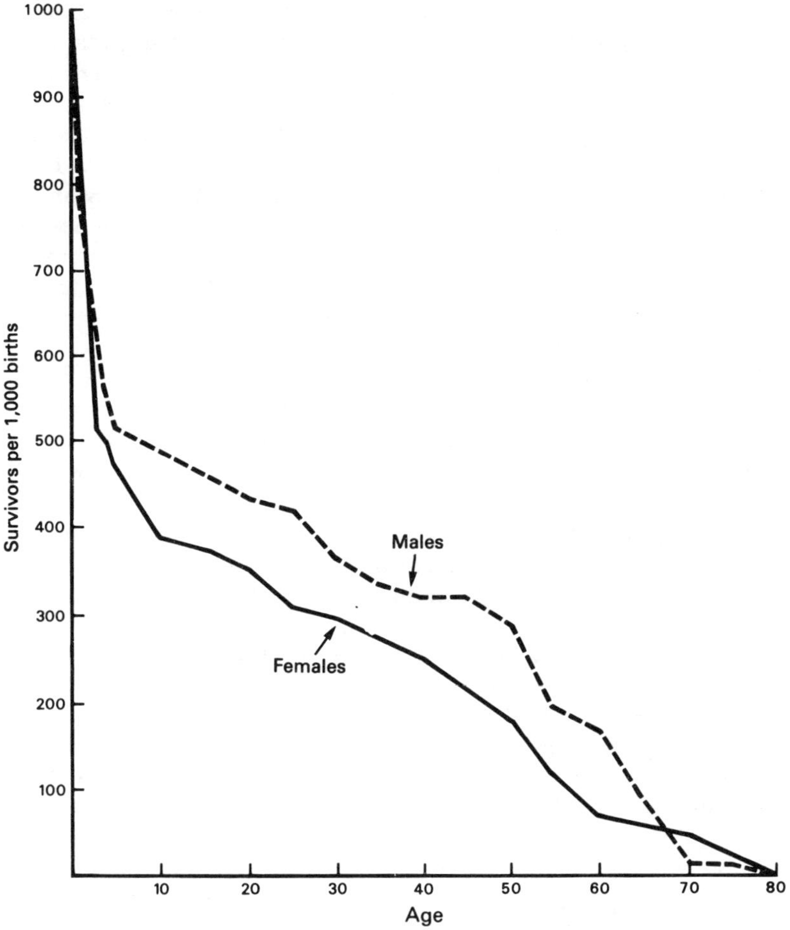

Source: Haswell (1975: 156).

household) up to one third of the value of the previous year's crop, a practice that had become synonymous with 'subsistence credit'. Groundnuts are lifted from October, wind-rowed and stacked on the haulm in the field. Threshing takes place in December/January when the trade season opens. The undecorticated (in shell) ground-nuts are stored in bags supplied by the co-operative, and transported,

155

mainly by donkey cart, from the field to the nearest co-operative buying centre (*secco*) one of which, as we have stated, is sited in Genieri. The co-operative buying agent tips the groundnuts from the bags to dry in the sun, raising one large heap of several hundred tons. After a few weeks, he hires labour to rebag the groundnuts and load them on to marketing board lorries for transportation to the Banjul processing plants. Poorer households were already heavily indebted to moneylenders, traders and relatives even before they had lifted their groundnuts, despite having received loans against their previous year's groundnut crop. While customary labour exchange had not completely disappeared, the practice, by now, of demanding money-wages in addition to food left many farmers who could not pay them unable to get labour.

Population and Change in the 1980s

The flight of immigrants, whole families upstaked in the wake of the severe droughts of the 1970s, left Genieri visibly scarred by vacant lots over which previous tenants had had no permanent tenurial rights. (Appendix: Table 8.5 indicates the sharp decline in the rate of growth of immigrant population.) Fifty per cent of the households had left, whole families moving to the periphery of the rapidly developing small town of Soma (Appendix: Table 8.5 and Map 8.2), to Farafenni town on the north bank, to Banjul on the coast. The demographic structure of the village had reverted to its original pattern of founding lineages. But the labour profile was now changed in those households that were sending children to the primary school in neighbouring Kaiaf village, first established by the Nutrition Field Working Party in 1948. These children were not available for customary bird-scaring activities. Two large households of Arabic teachers did not send children to school; their resident male pupils contributed labour in exchange for tutoring. In 1987 37.5 per cent of the smaller number of households contained more than 20 persons (an average of 36); these were almost exclusively founding lineage households which have retained their traditional power-base and extended family links, and in which the 'wealth' of the village had been concentrated a decade earlier when animal power had become an enabling technology. However, the effect of recent droughts was reflected in changes in the usage of animal power both for dry land cultivation and for the transport from field to village of harvested crop.

Competition for domestic water between the human and animal

population (including horses, oxen and donkeys) had resulted in an almost universal switch from oxen, with their much higher water requirements, to donkeys; but donkeys can only break the 'scarcity of season' constraint at planting time, while oxen can be used for the heavier work of land preparation, which generally leads to higher yields. Furthermore, the discovery by scientists of the presence of aflatoxins in foods and feedingstuffs had given rise to important questions concerning the risk to animal and human health posed by these mould-produced toxins: mould contamination and aflatoxin infection in the groundnut crop had become a prime concern of The Gambia's producers and exporters.

The conditions for growth of *A. flavus* and aflatoxin production depend on the moisture content and the nature of the substrata, the temperature and the available oxygen. Landell Mills Associates, in a study undertaken for The Gambia Government,[16] state that the period of greatest sensitivity of the groundnut plant to drought

> occurs about six to eight weeks after sowing, this being the period of vigorous flowering. However, drought after the pegging stage can also affect quality as far as aflatoxin contamination is concerned. When rain follows on protracted drought, some groundnut pods split in the ground. It would appear that the pods do not resume their growth as rapidly as that of the kernels, with the result that the pods split. Growth of moulds, including *A. flavus*, generally occurs along the sutures and below the partially exposed kernels of the split pods. . . . During the curing the kernel moisture content of groundnuts is generally in the range of 14–24 per cent; interruption and retardation of the field drying cycle by showers, dew or overcast humid weather, or a re-uptake of moisture after threshing and storage, usually results in the development of *A. flavus* with subsequent toxin formation or the development of 'black nuts' due to infection with *Macrophomina phaseoli*.

One of the main deterrents rested on the ability of farm households to 'deliver early' to their local buying points; and, by definition, such crops were those that had been early planted, early weeded

16. Grain Storage, Crop Losses and Groundnut Marketing Study undertaken for the Government of The Gambia (1981: 19, 21). Insect pests in cereals in household stores did not appear to present a serious problem except in threshed grain which was inadequately protected from insect attack. More serious was rodent infestation, and households with insufficient foodgrain to carry them through the long dry season could ill afford to feed rats. Better protection was found in households of a small tribe, the Balanta, that had settled about 20 km east of Soma (see also van der Ploeg, this volume, Chapter 6). The Balanta stored threshed grain in hermetically sealed earthenware pots made of mud plaster bonded with rice straw.

and early harvested. Farm households depended increasingly on the 'donkey-package' for these cultural operations; but, still debilitated at the beginning of the rains by poor feeding during the dry season, the death of a donkey during fieldwork was not uncommon.

As disturbing as the recent droughts was the abandonment of much of the swamp because of salt encroachments. Women were now far from meeting family subsistence needs from the production of rice. This raises the question as to whether 'dry' land farming on the guinea savannah soils alone could support the village population (by 1987 exceeding 1,000); this would require not only changes in technology, work-sharing roles and the pattern of crop production, but also in customary land tenure and rights of usufruct and access to credit.

A 1987 review of the economic status of the community forty years after the NFWP first made surveys[17] indicated that not all cultivable guinea savannah soils were actually being cultivated, that extensive new 'bush' encroachments would require male group labour to cut, burn (presently banned and carrying a heavy fine) and remove stumps; this labour was no longer available for this task because of the proliferation of small households (an average of 13.2 persons in households of 20 persons and under). Discussions with Genieri farmers revealed that since 1985 a 'European' had been given the usufruct of a large area, which on investigation proved to be land tractor-ploughed and strip-cropped by the NFWP in 1948 (Map 8.1, West C); the 1962 resurvey had found this plot fully cultivated by several farmers using traditional hand-tool methods. 'It is a big farm and we cannot cultivate it ourselves,' they said. 'Mr Tom came with a tractor and we lent him the land without any payment and he planted maize. If we have a tractor or similar mechanical equipment to plough our own land we will take that land back from him and do the planting for ourselves.' When interviewed, the 'European' said the land had not been used for seven years before his arrival and was covered with tall grasses. He agreed he paid nothing for the use of the land, and added that he had sold his 1986 maize through the statutory marketing board, was dissatisfied with the price, and would in 1987 be selling on the open market where, in all probability, its final destination would be over the border into Senegal.

It would appear that Genieri farmers have been overtaken by environmental changes; that they have not adapted fast enough to

17. My 1987 review of population and change over the past forty years in the village of Genieri and its environs, was supported by a Nuffield Foundation grant.

ensure sustained agricultural growth, for example, by extending animal-power cultivation to women for 'dry' land millet production (though there may be objection to carrying babies while ploughing); or by diversifying into vegetable production to supply Soma's non-agricultural population (Appendix: Table 8.6). In the present crisis over food security women try to market produce from their tiny gardens within the fenced area of their compounds. 'We travel to Soma but if you meet with people having the same basket of pepper it is very hard to sell. You may not even get enough money to buy a cup of rice, and sometimes you return empty-handed not even able to pay for the 'bus and bring back what you have not sold which afterwards "spoil" and are thrown away' (Appendix: Table 8.12).

Soma's growth and economic activity is underpinned by its primary transport routes, the surfaced road from Banjul the capital to Mansa Konko District Headquarters and the trans-Gambia highway (Map 8.2). In 1947 Mansa Konko was a bush-covered, uninhabited laterite rock, and to reach the District Commissioner a canoe trip across the river to the north bank was necessary; the trans-Gambia highway had not been built. With the construction of an all-weather road there have been highway-related public sector improvements, including immunisation of children and district maternity clinics. Van Dusseldorp (1971) classifies types of centres, assuming concentration is the starting point for the policy of establishing social and economic services in rural areas, given fast and efficient road networks:[18] see Table 8.4.

In the Van Dusseldorp model, the secondary centre includes shops, a small market, a workshop/pump station, storage facilities, rural industries (grain mills), primary and secondary schools, combined health centre and hospital, a postal agency, piped water, electricity supply, better housing and better sanitation. Soma, with a population in 1983 already approaching that of a small town (Appendix: Table 8.6), now falls well within this type of centre, and provides to a greater or lesser extent all the above services. Soma today is a rapidly developing urban growth pole. Land at fringe locations within Van Dusseldorp's (1971) small town radius of action of 8–20 km may be expected to become valuable for building if not to give rise to a speculative land market. Boserup (1965) warns:

When population growth becomes so rapid, in a pre-industrial

18. D. Van Dusseldorp, *Planning of Service Centres in Rural Areas of Developing Countries*, Wageningen, 1971, p. 131.

Map 8.2 Population catchment area, Soma, 1983

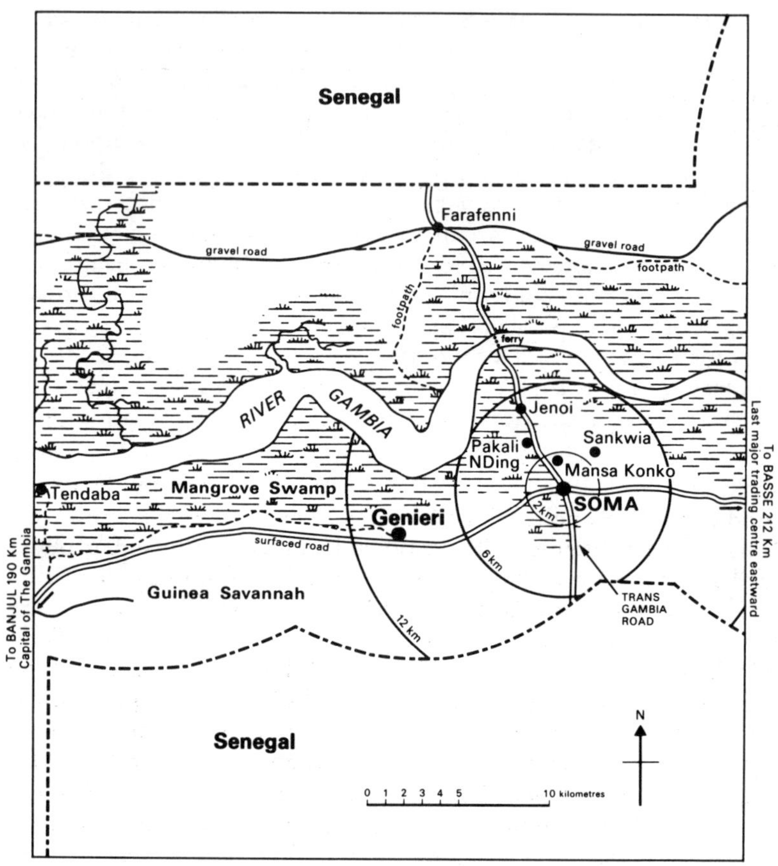

economy, new and difficult problems, unknown under slow or moderate growth, present themselves. The cultivators must be able to adapt themselves quickly to methods which are new to them, although they may have been used for millenia in other parts of the world, and – perhaps even more difficult – they must get accustomed, within a relatively short period, to regular, hard work instead of a more leisurely life with long periods of seasonal idleness. Moreover, the community must somehow be able to bear the burden of a high rate of investment and perhaps to undertake sweeping changes in land tenure.[19]

19. E. Boserup, *The Conditions of Agricultural Growth*, London, 1965, p. 64.

Table 8.4 The Van Dusseldorp model

Type of centre	Other names often used	Size of population	Radius of action
Additional primary	Hamlet	1,500	2 km
Primary	Local centre, village	1,500–5,000	3–7 km
Secondary	Large village, small town	5,000–10,000	8–20 km
Tertiary	Town, city	over 10,000	over 20 km

That the price of agricultural land is a consequence of its rent is not a meaningful concept under the traditional system of usufruct still extant in Genieri (though the 1987 review identified one woman, the first woman to become head of a household, who had purchased her household plot, paid for with a cow and two calves). Even so, the evidence pointed to a society which was beginning to modify its institutions within the wider perspective of encroaching 'rural–urban' settlement. The women had formed pressure groups, and in 1987 were calling for grain mills, an electricity generator and, from the men, land over which they had no 'rights', for vegetable production. However, while the population 'pull-effect' of transport networks linking Soma had dramatically changed the demographic structure of Genieri households from large to small units, this had not been matched by social change in response to the environmental crisis within the village caused by recent droughts.

In part, this may have been a resistance to change by traditional family household heads; 12.4 per cent of the population, first appearing in the 1949 census, had never left the village. In part, this may be a reluctance of male farmers to treat farming as a business. 'We have a good system of marketing our groundnuts nowadays,' they said.

We have animal-drawn carts for local transport, community seednut stores, seed-dusting against pests and diseases, fertilisers, and our own or hired animal-drawn implements for planting and weeding; but everything has a price. Nowadays you have a lot of money but it is not as blessed as before because, with your expenses, today's one thousand dalasis will not give you even what yesterday's one hundred dalasis gave you.

To sum up, forty years ago the men of the household worked

together to cut and burn a 'bush' farm for late millet. These farms were some distance from the village on guinea savannah soils which produced total crop failures after four or five years' cropping, and required a long period of bush regeneration before they were considered worth bringing under the hoe again. Those households that also grew early millet did so adjacent to their compounds, largely because it is not an awned variety and is subject to attack from birds. Children spent many hours scaring birds and monkeys, both in the distant fields of late millet, and in protecting the early millet. Children also went to the rice fields to scare birds when the crop was ripening; and boys went to the fields bordering the bush to scare monkeys off the men's individual plots of groundnuts grown for cash, in addition to spending many hours herding sheep and goats.

The women, working in 'kitchen' groups, farmed entirely for food production. They spent an exorbitant number of hours in preparing upland rainfed rice land to catch the early crop and in their flood-irrigated swamp rice, which, though more productive, could not be double-cropped because the river is tidal and salt during several months of the year. Once again, children spent many hours scaring birds off the ripening crop.

As a means of combating the disadvantages of a limited growing season, labour groups within the social structure were frequently employed. Notable among these were the age-sex group organisations, and the women's groups; these latter at times ran a communal farm as a hedge against famine; and because women had authority at certain ceremonies and were respected, their authority spread over other areas. In times of extreme need they gave assistance to one another.

This subsistence, or home-consumption, economy was suddenly overwhelmed with European contacts, bringing not simply new techniques affecting farming, but also the promise of extensive educational and other services with the opening in 1948 of a new administrative centre at nearby Mansa Konko.

Today, forty years on, Genieri farmers can be said to have moved a considerable distance from the subsistence end of the spectrum towards the commercial end of the scale. What have been the catalysts on the way? A generation earlier, population growth was virtually stationary, if not declining, in part as a result of high infant mortality rates, often associated with the strenuous agricultural activities of the women who bear and rear the children at the same time as producing the main food crop, rice; in part as a result of chronic malnutrition, and the incidence of malaria and other

162

diseases carried by insect vectors. The pattern of disease was dominated by both bacterial and viral infections. Yet by the late 1980s, the village was experiencing a high rate of population growth. Primarily, the catalyst was the construction of a surfaced road bringing motorised transport for quicker and more efficient evacuation of the cash-crop; road and transport networks also made more accessible to the community a health centre and maternity clinic established at Mansa Konko administrative headquarters. The demographic structure too has changed dramatically. Few large households of extended family members remain; and these are families from founding lineages, families that were highly regarded in the past and have risen on the wealth and prestige scale as the years went by, households with overwhelming resources in human energy applied to favourable resources in land. These are families that still own draught oxen for the more human-energy demanding tasks, and that have the resources to hire labour for some agricultural operations while their family members engage in off-farm employment and trading – lead families of the past who are the big men of today, continually adapting to successive change with new combinations of family enterprise. And it is the women of these households who are most vociferous in their demands for labour-saving technologies. For although the proximity of the village to a surfaced highway has been a major source of change, rice, the main farming activity, is still operated as a subsistence activity. Nowadays, this is severely constrained by the absence of children in educational institutions, and, in that majority of households today containing fewer than twenty persons, by the fewer 'kitchen' work groups and greater difficulty women find in getting reciprocal labour at seasonal peaks. But the constraint of time and energy-demanding tasks of hand-pounding grain is overriding.

Nancy Eisener asks 'which milling system?' In a simple handbook, she introduces the machine that dehulls and grinds sorghum. She prefaces this by giving a processing time of 3–4 hours for soaking, hand-pounding and winnowing 10 kg of grain, compared with 2 minutes when mechanically dehulling and grinding the dehulled grain into meal. The consumer-service milling is identified as the simplest type:

> people bring their own grain to the mill for processing. They pay a milling fee based on the total number of kilograms milled. . . . Storage requirements are minimal. . . . Fuel for the engine and customers for the service are the only things needed for the daily operations. . . . A consumer service mill has a low capital cost and it could be in any size of

village. There is no product marketing necessary as the customers supply their own grain and take away the meal and bran.[20]

And while she makes the obvious statement that grain is essential for the success of the mill, she does add a rider, that tailoring its products to consumer preference is of central importance if people are to be attracted to its use. This calls for a better understanding of people's tastes (see Chapter 11, p. 237). A sorghum milling feasibility study carried out in Tanzania indicated, however, that to support a small, diesel-operated dehuller with an 8–hour/day throughput of 2,000 kg, would require an annual foodgrain production of around 672 metric tons, a production which even a typical Tanzanian settlement of about 600 families with an average of seven household members could not attain without growing much higher yielding, more drought-resistant, introduced sorghums. Although the present population of Genieri village would not support a like mill even if there had been a switch in the pattern of production from wet to dryland farming as salt encroachment forces women off the swamps, the rapid development of Soma does warrant the introduction of sorghum dehullers in and around the township. One important feature of the Tanzanian study was, however, that it identified a village community which was producing large marketable surpluses of sorghum for dehulling, but could thresh only half the crop because of an acute shortage of family labour. This warns of the difficulty of altering one part of a production system without changing other parts: a sorghum threshing machine may be a useful component of any package which includes dehullers at the township level.[21]

A sustained rate of population growth with better medical services, and a higher population density, will be tomorrow's catalysts for technological innovation and change in Genieri village; and there are sociological aspects which may be expected to follow. Women, already used to group activity, will demand the same facilities of a cooperative as are now enjoyed by the men. Young children will be put into day-care centres, with older children in

20. This handbook ('The Machine that Dehulls and Grinds Grain for Us') is well illustrated, and contains useful appendices on grain processing records, extraction rates, and so on.

21. This study was carried out in March 1983 under the Commonwealth Fund for Technical Cooperation, by the Tanzanian Ministry of Agriculture, the Tanzanian Small Industries Development Organisation, and Margaret Haswell Associates, whose brief was to provide countrywide information on the market for village dehullers.

school and not free to look after them. Women will extend their farming activities into dry-season cultivation of vegetables provided that the water-table permits, and they will share some crop and livestock production tasks currently performed by the men as the men devote more time to off-farm employment. For yesterday's luxuries are today's necessities in an economy which is fast becoming monetised with the proximity of a rural–urban growth pole.

While the village appears to be experiencing a period of economic stress at the present time, largely caused by the community's tardiness in adapting to environmental changes, the comparatively dense population of rapidly developing Soma township, which is absorbing a share of their labour and will depend increasingly on a regular transportation of food, will eventually lead to sweeping changes in land tenure and in patterns of agricultural production.

Appendix

Table 8.5 Annual population growth rate at successive Genieri village censuses[a]

Year	Total	Average annual growth rate	Descendants of original settlers	Recent immigrants
1949	483	. . .	. . .	. . .
1962	587	1.6	–0.5	5.1
1973	764	2.4	1.0	3.6
1982	874	1.5	2.4	–1.1

Source: Successive censuses taken by Margaret Haswell and Alhaji Ebrima M. Fye.

Table 8.6 Mansa Konko Local Government Area[b]: Growth of 'small town' population

Type of centre	Year	Total	Average annual growth rate
Soma township situated on trans-Gambia highway between Cassamance and Senegal	1973	1,267	. . .
	1983	4,788	14.2
Outlying villages < 2 km from Soma township	1973	3,602	. . .
	1983	4,685	2.7
All enumerated village populations < 6 km from Soma township	1973	7,862	. . .
	1983	10,600	3.0

Source: National Statistics, Ministry of Economic Planning and Industrial Development, Department of Statistics.
a. Genieri village lies within the Mansa Konko Local Government Administration, Lower River Division, The Gambia (Map 8.2).
b. The 1983 National Census proportion of males to females in the Mansa Konko Local Government Area is recorded as 48.8 per cent compared with 51.4 per cent in Banjul the capital and its urban agglomerations on the coast in which more than a fifth of The Gambia's total population reside.

Table 8.7 Land and labour use for crops, Genieri, 1949

	Total area ha	Average hours labour per ha (all operations)
	Men's work	
Guinea savannah		
Late millet	35.8	544
Early millet	5.7	660
Sorghum	4.1	166
Maize	2.3	786
Groundnuts	105.8	650
	Women's work	
Digitaria exilis 'famine' food	32.6	116

	Women's work	
Rice flats		
Upland rainfed rice	28.0	2,051
Swamp flood-irrigated rice:		
seedbeds	9.3	1,534
transplanted	69.7	(based on transplanted area)

Source: Haswell (1953: 32).

Table 8.8 Ratios of energy gained to energy expended

	Including walking to and from farm plots	Per cent of total human labour energy costs	Per cent of total area cultivated
Late millet	8.1	7.2	12.0
Early millet	12.4	0.9	1.8
Sorghum	13.9	0.4	1.7
Maize	13.4	0.8	0.8
Groundnuts	14.7	28.1	35.5
Digitaria exilis	16.8	1.3	11.3
Upland and swamp rice	5.3	61.3	36.9
Net energy surplus overall	8.4		

Source: Haswell (1985: 28). Summary of the ratios of energy gained to energy expended in subsistence farming, Genieri, 1949–50.

Table 8.9 Age and sex distribution of farmers in Genieri village 1949–50

Age	Males %	Females %	Both sexes %
Under 11	4.3	—	4.3
11–15	8.4	5.6	14.0
16–19	7.4	4.0	11.4
20–39	19.2	31.9	51.1
40–49	6.8	5.6	12.4
50 and over	2.8	4.0	6.8

Source: Haswell (1985: 29).
Note: Girls under 11 years of age did not farm.

Table 8.10 Population and area cultivated by Genieri farmers

	1949–50	1961–2	1973–4
Population	483	587	764
Cultivated area:		hectares	
River flats			
Upland rainfed rice	27	31	22
Swamp flood-irrigated rice (including seedbeds)	77	118	139
Guinea savannah			
All other crops	176	142	205
Total cultivated area	280	291	366
Cultivated area:			
per household	14.7	8.1	7.3
per head of population	0.58	0.50	0.47

Source: Haswell (1975: 162).

Table 8.11 Average (mean) yields per hectare: six recorded households, Genieri, 1973–4

Crop	Method of cultivation	Average size of plot	Yield	
		ha	kg/ha	
Late millet	Ox-ploughed	2.01	527	grain
	Hand-cultivated	0.76	329	
Early millet	Not grown by sample households[a]			
Sorghum	Ox-ploughed	1.44	109	"
Digitaria exilis	Hand-cultivated	0.14	681	"
Groundnuts	Ox-ploughed, planted by donkey-drawn mechanical planter, and chemical fertiliser applied[b]	1.54	1,570	nuts in shell
	Ox-ploughed, hand-planted		1,005	"
	Hand-hoed, planted by donkey-drawn mechanical planter	1.16	1,163	"
	All operations by hand		893	"

Upland rainfed rice	Hand-cultivated	0.08	676	paddy
Swamp flood-irrigated rice[c]	Hand-cultivated	0.37	1,506	paddy

Source: Haswell (1975: 168).

a. Only four households grew early millet
   (8 per cent of the village households).
b. Superphosphate purchased from Senegal.
c. The women of the household work in kitchens. In extended family households
   each wife has her own kitchen and her own rice store, and each has her own plots
   on which she is assisted by younger members of her immediate family. The
   average number of plots cultivated per *kitchen* was 2.5.

Table 8.12　The contribution of cash receipts to total family income in
rice equivalents, Genieri village, 1987[a]

	Per cent person/year
Receipts from sale of cash crop groundnuts	23.3
Receipts from sale of vegetables in Soma market	1.4
Receipts from sale of firewood	0.5
Household earnings from non-agricultural occupations: tailor, mason, carpenter, fisherman, native doctor	2.4
Wages for public services: President of the Cooperative, watchman, policeman, labour at groundnut buying station (Genieri serves a wide area of outlying villages)	2.5
Earnings of arabic teachers living in the village	1.1
Remittances from relatives living elsewhere in The Gambia or abroad	6.2
Food aid (rice in August distributed to every household)	4.5

a. Based on a 'minimum' foodgrain requirement of 160 kg per person per year.
   Cash receipts have been converted into rice equivalents at the local price of clean
   rice: (Dalasis) D.2.00 = 1 kg. For food security, some two-thirds of home
   consumption needs must be produced on the farm (considerably more if grain is
   fed to draught animals). The rapid rate of growth of the non-agricultural
   population in Soma new town finds outlying villages becoming market-
   orientated in respect of all foodgrains, fruits and vegetables, but without organised markets for horticultural produce, Genieri women were often unable to sell
   their small surpluses. *This calls for early review of agricultural pricing policy and
   marketing structures to stimulate the production of marketable surpluses.*

Table 8.13 List of crops grown

Crop	Botanical name	Local name
Late millet	*Pennisetum typhoides* var.	Sanyo
Early millet	*Pennisetum typhoides* var.	Suno
Guineacorn	*Sorghum vulgare*	Basso
Red Basso	Sorghum spp.	Kinto
Groundnuts	*Archis hypogea*	Tio Messeng (spreading variety)
Rice	*Oryza sativa*	Mano
Findi	*Digitaria exilis*	Findo

Small quantities of the following are planted
around the household compound

Crop	Botanical name	Local name
Maize	*Zea Mays*	Tubab Nyor
Cassava	*Manihot utilissima*	Nyambo
Okra	*Hibiscus esculentus*	Kanjo
Pumpkin	*Cucurbita maxima*	Jenghor
Bottle-gourd	*Lagenaria vulgaris*	Bolo
Pepper	*Capsicum grossum*	Biddulph Kano
Tomato	*Lycopersicum esculentum.*	Tubabo mentengo
	Solanaceae	Mentengo

References

Boserup, E. *The Conditions of Agricultural Growth: The economics of agrarian change under population pressure*, London, 1965

Clark, C., *Population Growth and Land Use*, London, Macmillan, 1967

—— and Haswell, M., *The Economics of Subsistence Agriculture*, 4th edn, London, Macmillan, 1970

Cummings, R. W., 'Expectations for Future Developments in Sorghums, Millets and Legumes', Paper presented at the conference on Science and Agribusiness in the Seventies, London, February 1974

Eastman, P., *An End to Pounding: A new mechanical flour milling system in use in Africa*, Ottawa, 1980

Eisener, N., *The Machine that Dehulls and Grinds Sorghum for Us*, Rural Industries Innovation Centre, Botswana, 1979

Gamble, D. P., *Economic Conditions in Two Mandinka Villages: Kerewan and Keneba*, Colonial Office, 1955

Government of The Republic of The Gambia Report No. MA33856/(24)/RDP, *Grain Storage, Crop Losses and Groundnut Marketing Study*, prepared by Landell Mills Associates, 1981, mimeo

Haswell, M. R., *Economics of Agriculture in a Savannah Village*, Colonial Research Study No. 8, HMSO, 1953

—— *The Changing Pattern of Economic Activity in a Gambia Village*, Department of Technical Co-operation Overseas Research Publication No. 2, HMSO, 1963

—— *The Nature of Poverty*, London, Macmillan, 1975

—— *Energy for Subsistence*, 2nd edn, London, Macmillan, 1985

Lawrence, P. and Smith, A., 'A Better Beast of Burden', *New Scientist*, 21 April 1988

Van Dusseldorp, D. B. W. M., *Planning of Service Centres in Rural Areas of Developing Countries*, International Institute for Land Reclamation and Improvement, Wageningen, 1971

9

The Impact of the Tobacco Industry on Rural Development and Farming Systems in Arua, Uganda

Josephine Wanja Harmsworth

This chapter is based on documentary sources and field surveys in the West Nile Region of Uganda, carried out in 1988–9 by a team of four researchers.[1] It is part of a broader study that was commissioned by a subsidiary of British American Tobacco, BAT Uganda (1984) Ltd, on the impact of the tobacco industry on the economy of Uganda. The field surveys were undertaken at a time when BAT was introducing new designs of furnace for tobacco curing kilns, and after considerable rehabilitation had taken place to both co-operative society and individual property used in the production and marketing of tobacco. West Nile was, however, still being disrupted by bandits, which made it dangerous to visit more remote northern communities. Some of the peasants, who had fled as a result of armed conflict in 1980–1, were only just being resettled. This situation necessarily impinged on results.

The field surveys included clustered samples of 500 farmers, 200 farmers' wives, and several tobacco and other co-operative societies and unions in four counties of Arua District. The farmers and their wives were both growers and non-growers; the co-operatives included both tobacco and cotton societies. Earlier fieldwork of basic importance includes that of Middleton in 1950,[2] Hutton in 1968,[3] and Mackenzie in 1971.[4]

1. The Arua field surveys were carried out by Professor J. Opio Odong and J. Fendru (Makerere Faculty of Agriculture), L. Msemakweli (Uganda Central Cooperative Alliance) and J. Harmsworth (Team Leader and Private Consultant), in July and September 1988.

2. J. Middleton, 'Trade and Markets among the Lugbara of Uganda', in P. Bohanan and G. Dalton (eds), *Markets in Africa*, Northwestern University Press, 1962.

3. R. C. Hutton, 'Unemployment and Labour Migration in Uganda', PhD

Map 9.1   Uganda north–western region

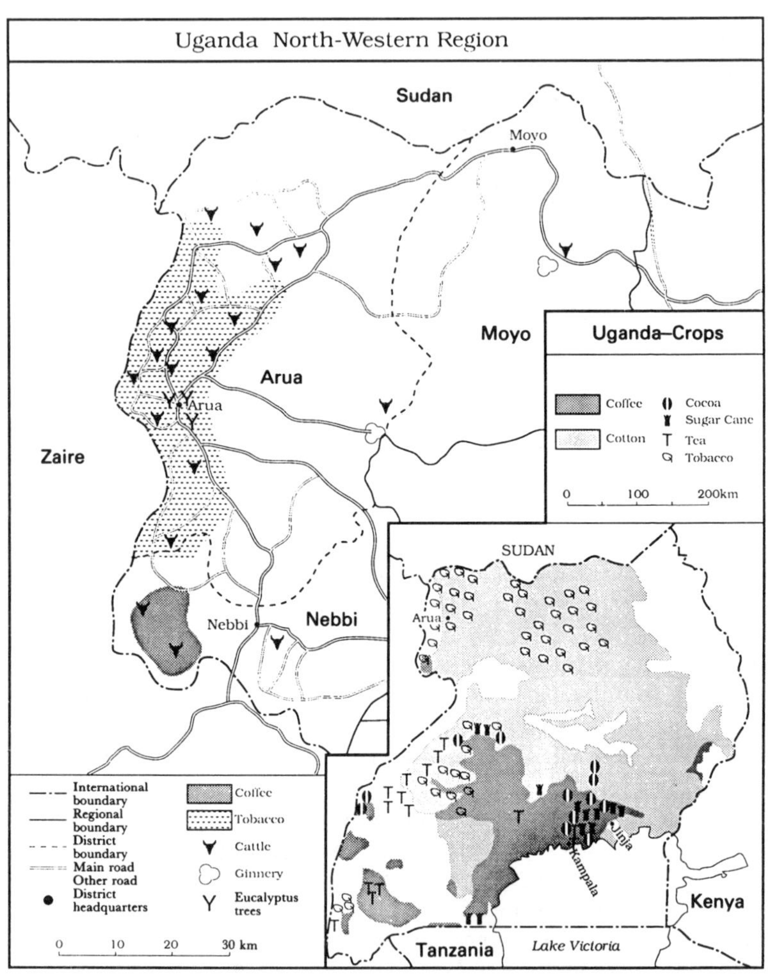

While the introduction of tobacco culture to Arua has inevitably influenced development in the region and has had a considerable impact on growers and their families, it is not always easy to isolate particular effects, and in most cases impossible to quantify them.

---

thesis, University of East Africa, 1968

4. M. K. Mackenzie, *Present Location and Past Diffusion of the Flue-cured Tobacco Industry in West Nile District*, Makerere University Geography Department, Occasional Paper No. 21, 1971.

Tobacco was introduced more than fifty years ago, in 1936. Various other agricultural innovations preceded tobacco and many other events, some of a revolutionary and even violent nature, occurred both before and after the introduction of tobacco. Since 1920 there has been a large outmigration to work in the south. Another major consideration is that the population has more than tripled in the same period.[5] Meanwhile the technology associated with tobacco growing has itself been subject to changes, some of a quite fundamental nature.

The initial forward momentum established up to 1970 was drastically undermined by the maladministration of the Amin era and subsequent social and political upheavals. Even now West Nile, the main flue-curing producer area in Uganda, is partly cut off from the rest of the country by continuing rebel activities. While 1987 and 1988 saw substantial increases in production, in 1989 the region was hit by a devastating drought and by cassava mosaic disease, which virtually wiped out the entire most important food staple. At the same time, due to a combination of bad road surfaces and banditry, the 300 mile journey to Kampala was taking up to a month or more.

The major national impact of the tobacco industry in Uganda has been through the manufacture and domestic marketing of cigarettes. Taxes on tobacco and cigarettes today form a substantial percentage of government revenues. While tobacco has never compared in export performance with the two main foreign exchange earners – coffee and cotton – nevertheless at its peak in 1972 it was one of a handful of more important diversified exports. BAT pays some of the highest salaries in the country, and therefore its staff contribute disproportionately to total PAYE. Company taxes and dividends also benefit the national treasury. The total of these taxes and dividends amounted in 1988 to the equivalent of 53 million US\$. (See Appendix: Table 9.17 for foreign exchange rates over time.)

There have also been considerable contributions to regional growth and development through the cultivation of leaf as a valuable cash crop, and through associated technological innovations and the incidental introduction of particular forms of market organisation. The area which has been most radically affected is Arua District, which presently accounts for about 70 per cent of national production. While the main thrust of change has been on

5. Uganda Government, Population Censuses 1948, 1959 and 1969; Background to the Budget 1988/9.

growers, who form a small minority of farmers in the whole of Arua but a higher proportion in the main grower areas, tobacco production has also had an impact on non-growers, on their communities and in fact the whole district.

These changes, when viewed over time, have been substantial, and can best be understood by reviewing what Arua was like in 1936. Up to that time there were no organised markets, and most households were almost completely self-sufficient.[6] Even as late as 1950 most women still wore the traditional dress of special bunches of leaves. Tobacco was partly responsible for bringing Arua into the monetary economy. It was also associated with advances in standards of living and of education. In so far as agriculture was concerned, it introduced new methods and ideas, and later on the technology associated with the curing process changed patterns of work and farm management.

The Historical Background

The tobacco-growing areas of Arua, which comprise the counties of Maracha, Ayivu, Terego, and parts of Vurra, Koboko and Aringa (see Map 9.1), are the most densely populated of West Nile and also the most fertile. They are inhabited by the Lugbara people, who are patriarchal, patrilocal and patrilineal. However, affines often have important functions. The Lugbara do not have clans of the Bantu type, but more localised kin-groupings. These kinship clusters, which have common ancestors, occupy contiguous land, although over time various non-kin may have come to live in the same places so that today not all the people of a village are related to each other. About one third of all Lugbara are in Zaire, and cross-border linkages are important. There is a tobacco factory in Zaire near the border and some Ugandan tobacco is smuggled to Zaire, and vice versa, depending on price incentives.

The typical Lugbara household consists of a man with his wife/ wives and children. This was described by Middleton some forty years ago,[7] and it was assumed that this nuclear unit was spatially integrated. However, recent research shows a large number of polygamous families with wives living in different homesteads.

The history of the Lugbara is one of migration and of assimila-

6. J. Middleton, *The Lugbara of Uganda*, Northwestern University/Holt, Rinehart & Winston, 1965, p. 25.
7. Ibid., p. 7.

Table 9.1   Rate of labour migration

Countries	Density of population per square mile	Labour migration rate: adult males (per cent)
Maracha	240	27
Ayivu	239	21
Terego	80	18
Vurra	65	14
Aringa	16	12

Source: Middleton (1960).
Note: One square mile equals 2.6 square kilometres.

tion with neighbouring people. By the nineteenth century they traded extensively with the Ndu who are ironworkers, and with the Alur who obtained salt from southern markets.[8] In so far as is known, they have always been agriculturalists with some livestock. West Nile first came under the control of Belgium at the end of the nineteenth century. The area was then briefly under the suzerainty of Sudan. In 1914 West Nile became part of Uganda.[9] Taxation was introduced but little was at first collected since the population had no source of money income. Missionaries and traders came to the area around 1917.

In the period between the two world wars change came gradually. West Nile is far from the centre of political power and commerce in Kampala and even up to now there are no tarmac roads except in Arua town. The region was first integrated into the mainstream of Ugandan life through outmigration to the plantations, factories and security forces in the south. This migration started in a more organised way around 1920 when recruitment was carried out for expatriate enterprises, but was accompanied by substantial spontaneous individual migration.[10] By the 1960s few Lugbara males had not at some time in their lives worked in other parts of Uganda. Figures for 1952 indicate that varying proportions of adult males were away from different parts of the district (see Table 9.1). This outmigration had very substantial political, social and economic implications.

8. J. Middleton, 'Trade and Markets among the Lugbara', pp. 562–4.
9. J. Middleton, 'Social Change among the Lugbara of Uganda', *Civilisation*, Bruxelles, vol. 10, no. 4, 1960, p. 448.
10. Ibid., p. 449.

As Middleton states, 'the absence of young men has serious repercussions . . . wives are left without husbands, girls have no one to marry and take too many casual lovers . . . many men . . . leave kinship statuses unfilled . . . young men . . . no longer obey the authority of their elders. They keep their earnings for their own purposes . . . and so arrange their own marriages'.[11] Migration has continued and in some cases has become permanent.

Various happenings have been important in the more recent history of West Nile including drought, locusts and the Second World War. However, the single most recent important event has been the advent and overthrow of President Idi Amin. Amin relied very substantially on soldiers from West Nile to overthrow Obote,[12] and during his time in power people from the area enjoyed unprecedented opportunities in trade and in employment. With little or no education many were appointed to positions of power in parastatals or in industries acquired from the departing Asians in 1972. They remained dominant in the army. It is evident from superficial observation that there was substantial economic development in West Nile in the period 1971–9 at a time when there was a general decline in the national economy and a particular drop in the production of leaf tobacco.

In 1979 Amin was overthrown. Although there was no immediate violence in Arua, this was followed by an attempt in late 1980 and 1981 to reinstate Amin's supporters, and resulted in wholesale destruction of buildings in the rural areas, considerable damage to Arua town,[13] and the flight of almost the entire population of Arua and Moyo Districts to neighbouring Zaire and Sudan. People only started to return in 1983; by 1988 returnees were still being assisted by the UNHCR. In 1989 the process was reversed and the area is now the recipient of Sudanese refugees, some of whom may claim relationship with indigenous Ugandans.

The Introduction of Tobacco

Commercial varieties of tobacco were introduced to Uganda in 1927 and to Arua in 1936. It was not the first cash-crop in the area:

11. Ibid., p. 450.

12. R. Denis Pain, 'Acholi and Nubians: economic forces and military employment', in D. P. Wiebe and P. C. Dodge (eds), *Beyond Crisis: Development issues in Uganda*, Makerere Institute of Social Research, 1987, p. 46.

13. P. C. Dodge, 'The West Nile Emergency', in P. C. Dodge and D. P. Wiebe (eds), *Crisis in Uganda: The breakdown of health services*, Oxford, 1985, p. 108.

cotton had been grown in some places for much longer and sunflower was also tried. However, tobacco was ideally suited to places at higher altitudes unsuitable for cotton.[14] Tobacco was not a new crop, it was an item of trade in the earliest recorded markets[15] and still continues to be sold in local markets in Arua and even over the border in Zaire.

BAT, which presently has a monopoly position, was involved from the outset, stationing an officer in the then Bunyoro District from 1928 to 1936.[16] It was also instrumental in both the introduction of tobacco to West Nile in 1936 and in the further development of the industry from that time on. Up to 1962 it was responsible for all operations in West Nile, as well as the curing and ultimate processing of the leaf into tobacco and cigarettes in Kampala and Jinja. After 1962 extension work was made the responsibility of the Department of Agriculture, and in 1969 marketing was taken over by the newly established Produce Marketing Board. In 1972 Amin nationalised the company and from then until its return in 1984 BAT ceased to have any direct interest in tobacco growing in Uganda. In the interim the company became a parastatal and a partner in a World Bank smallholder tobacco project. This was initiated in 1971 and was supposed to double production, but it failed disastrously, although financial accounts were still being maintained up to 1984.[17]

The Process of Production

Tobacco production requires the co-ordination or integration of a number of activities. Its value depends at all stages on careful handling and maintenance of specific quality guidelines. Tobacco leaf, for instance, is graded and the price differential between top and lower grades can be considerable.

Tobacco is grown in seedbeds and transplanted. It must be carefully weeded. The use of various agrochemicals, fertilisers and pesticides, are essential if maximum yield and quality are to be achieved. When harvested, tobacco leaf has to be cured: either in

14. Mackenzie, *Present Location and Past Diffusion of the Flue-cured Tobacco Industry*, pp. 18, 29.

15. J. W. Purseglove, *Tobacco in Uganda*, Kampala, 1951.

16. Mackenzi, *Present Location and Past Diffusion of the Flue-cured Tobacco Industry*, pp. 47–8.

17. World Bank Project Completion Report on the Uganda Smallholder Project, IBRD, 1982.

flue-curing barns, in fire-curing barns, by air-curing or by sun-curing. All commercial production in Arua today is flue-cured, although traditional varieties are still sun-cured.

Fire-curing is a relatively simple operation. Flue-curing is more complex and requires greater capital outlay. Permanent barns have to be constructed which are roofed with thatch, iron sheets or tiles. In Uganda burnt bricks are used for barn construction. In addition to the capital cost of the barn, recurrent inputs include replacement of pipes, poles and twine. A metal flue is built into the bottom of the barn which draws flames from furnaces which are set into the walls of the barn. The leaf is tied in bunches on poles, which are arranged in tiers across the barn. Ugandan tobacco barns are usually 4 × 4 m × 6 tiers (about 7 metres high). Great care has to be taken to maintain consistent temperature levels until the tobacco has been cured. This is a 24-hour process, and it takes approximately four days to complete. The curing process affects not only the quality but various other properties of the leaf.

Flue-curing can require large quantities of fuelwood, and because of this the cultivation of tobacco has to be accompanied by the cultivation of trees and provision of transport for the wood. The cured leaf also requires to be transported to marketing centres preferably in hessian squares. There it is graded, weighed and packed. The leaf must then be dried again and threshed in a large-scale plant before it can be manufactured into cigarettes. The smooth flow of production relies on the ready and timely availability of inputs at every stage.

The organisation of tobacco production varies in different countries. In Uganda the growers are all small-scale farmers, and farmers' co-operatives are the intermediary in the purchasing process, while the manufacture of tobacco is currently the monopoly of one company which also handles exports and imports. An important feature of the present structure of the industry is the role played by co-operative societies, not only in marketing the crop but in maintaining fuelwood plantations. Since 1984 the Arua District Tobacco Union has also manufactured pipes and roof tiles for the barns.

In the 1960s BAT developed a model system for both extension and marketing of tobacco and since their return in 1984 have again reintroduced many of its features, which include provision of inputs on credit, extension advice, forestry extension and credits, and centralised marketing. Not only individual farmers but the co-operative societies and union have access to long-term, medium-term and short-term interest-free loans from the company. The company has also invested in other benefits to growers, such as the supply of bicycles at cost price and educational facilities.

The cultivation of tobacco was adopted readily by farmers in Arua, although standards of cultivation and curing are still below the optimum. More farmers were in fact prepared to grow tobacco than could be absorbed by the industry, which controlled and restricted expansion. The major concern in colonial times was to sustain purity of strain and leaf quality. Statutory instruments regulated where tobacco could be grown and marketed, and up until the late 1960s a quota system operated in Arua. Barn sharing was an attempt to keep the number of operative farmers high while stabilising output, thus maintaining local interest in the crop. Tobacco is cultivated by only a minority of farmers, concentrated in specific growing areas which are also limited. In 1988, 8,500 out of a total estimated 120,000 farmers in Arua were growers. Today expansion in numbers of growers is restricted by access to barns and limitations in financial and other resources necessary for their rehabilitation or extension. Another consideration is the need to plan for growth which is matched by capability to buy and handle the crop. An unexpected, indeed dramatic, increase of about 60 per cent in output of leaf from Arua in 1988 nearly resulted in cash-flow constraints for the company.

The Impact on Regional Development

The impact of tobacco culture on the development of Arua District in general and the farm population specifically can be discussed under several headings. These include:

1   impact on district economy;
2   co-operative development;
3   economic impact on farmers;
4   impact on farming practices and management;
5   social impact on farmers and their families, especially women.

Tobacco has made a major contribution to the development of a market-oriented economy as a result of injections of money into the system through purchases of goods and services by growers. By the time tobacco was introduced, and even up to 1950, the volume of trade was small. Local markets were few and functioned erratically. There is some evidence that not only trading centres with shops, but the development of regular markets took place largely after the introduction of the crop.[18] Growers purchased

18. Hutton, *Unemployment and Labour Migration*, p. 202.

exotic goods, such as bicycles, clothes, household utensils and farm equipment. Some of their wants were satisfied from expanded local production of such items as beds and chairs, thus contributing to secondary economic growth. Even today the contribution of total tobacco earnings in the district is very significant. In 1988–9 a total of Shs 373,973,710 was paid to growers.

Tobacco growers have also created employment by hiring labour not only for tobacco but for other crops. By 1968 45 per cent of growers hired wage labour.[19] In 1988 nearly all growers used hired labourers (see Table 9.2). The amount spent on labour is quite considerable in local terms as is shown in Table 9.3. In addition the curing of tobacco leaf requires a great deal of timber and this has necessitated the establishment of forest plantations. These also require labour to maintain. The need for transport for woodfuel and for the processed tobacco to Kampala, and of inputs from Kampala to Arua, also contributed to the job market.

Tobacco has required the adoption of novel technologies and the learning of new skills. Some of these have been applied in other spheres. The square-built house is copied from the style of barns, and even by 1968 Hutton reports that most compounds already had one.[20] The technology associated with flue-curing has also provided the impetus for much brickmaking. Bricks are today commonly used in individual houses. There are also one or two bakeries in Arua, which utilise skills learned with tobacco furnaces. As is discussed below under the heading of farming practices, there has in addition been some transfer of new methods of cultivation to other crops.

Another area in which interventions associated with the introduction of tobacco have had an impact on the more general development of the district is transport and communications. At one time the company was actively involved in construction of feeder roads necessary to reach farmers. This infrastructure is still in place, and has been important in opening up more remote areas to trade and development. Growers were among the first to own bicycles in Arua. These revolutionised the transport system in earlier years and are still a vital element in communications. Today the bicycle is the main form of transport for people and produce within the district, since regular buses or taxis are almost non-existent and very expensive. It also continues to be the major means of taking tobacco to marketing centres. As a result of company initiatives to

19. Ibid., p. 203.
20. Ibid., p. 170.

Table 9.2 Use of hired labour: per cent of farmers using on tobacco and food-crops

Crop	Growers	Non-growers
Tobacco	90	0
Cassava	70	46
Legumes	67	47
Simsim	65	36
Millet	58	39
Maize	43	32
Sweet potatoes	36	17
Exotic vegetables	10	11

Table 9.3 Amount spent on labour on tobacco 1988 in Ugandan shillings[a]

	Per cent growers
None	12
5,000 and less	26
5,001–20,000	55
20,001 +	5
No reply	2

a. Compared to the minimum monthly wage at that time of Shs 1,500. For foreign exchange rates, see Appendix: Table 9.17.

bring bicycles to the union for sale to growers, they still own more than non-growers. In 1988, 53 per cent of growers had bicycles and 10 per cent two and more, compared with 28 per cent and 3 per cent of non-growers.

The Impact on the Development of Cooperatives

Tobacco growing has necessitated new forms of organisation of family labour for cultivation and curing, and new forms of community co-operation for curing, marketing and the maintenance of fuelwood plantations. The latter has involved the formation and development of growers' co-operative societies. After an unsuccessful attempt to use co-operatives to cure the crop, they were reorganised in 1964 both to manage fuelwood plantations and to market the leaf.

Co-operatives are important as an intermediary between the company or government and the farmer, not only for provision of inputs and purchase of the cured leaf, but in mediating policy issues. At the present time, due to the shortage of fuelwood for curing, and the difficulties of obtaining and transporting iron sheets for roofing the barns, the company has introduced a new design of fuel-saving furnace, and provided assistance to the co-operative union to produce tiles for roofing the barns. These innovations have been discussed at all stages with the farmers through their societies. As a result action has been taken quickly to modify the flue-pipes and to improve on the standards of roofing arising from farmers' complaints about weaknesses in both. The union also participates in decisions made by government on pricing and other matters.

Tables 9.4 and 9.5 show the growth in importance of co-operatives, and in particular the tobacco societies, in Arua up to 1970. After 1971 co-operative activities declined, together with the drop in output of cash-crops. During the civil strife of 1980–1 in Arua there was extensive destruction of co-operative property. By 1988 only the tobacco societies and union were thriving, in spite of a special project to revive the cotton industry which is the other major alternative cash-crop in Arua, and other initiatives to re-habilitate coffee. Whereas the tobacco union and society stores and offices have been rehabilitated and even extended, and the union is now registering large net profits (Shs 16,765,052 in 1988) the cotton societies are without any resources, and work on rehabilitation of the cotton union ginnery at Rhino Camp, on-going since 1978, has only just been completed.

In 1988, the district co-operative office of the cotton societies had no vehicle and their offices were only sparsely furnished. Officials did not know how many active societies there were. They were therefore hardly in a position to provide any support. The tobacco societies and union, as a result of the continued and close relationship with the company, which includes the posting of agricultural staff at each and weekly visits by senior staff, had very active offices with display boards, well-kept stores with a wide range of needed inputs including tools, agrochemicals, building materials and specialised inputs like barn thermometers; and at the time of the survey were very active in crop buying. Up-to-date records also showed the extent to which farmers had been supplied with these inputs. The cotton societies, on the other hand, were completely inactive, and officials often difficult to locate. Furthermore, the problems undermining the effectiveness of other marketing co-operatives

184

Table 9.4 Total turnover of tobacco, coffee and cotton co-operative societies in Arua in 000s Ugandan shillings and percentage contribution by crop

Year	Turnover	Coffee %	Cotton %	Tobacco %
1952–53	67	—	—	—
1958–59	1,751	—	99	1
1967–68	12,373	22	68	10
1968–69	29,220	18	60	22
1969–70	29,171	25	53	22

Source: Mwaka (1978).
Notes: 1952–67: 1 Ugandan shilling equalled 14 US cents.
1967–70: 1 Ugandan shilling equalled 12 US cents.

Table 9.5 Percentage of total national tobacco production handled by the Arua co-operatives and value 000's Ugandan shillings

Year	Turnover	Per cent national co-op total
1958–9	12	1
1967–8	5,822	79

Source: Mwaka (1978).
Note: See note to Table 9.4.

nationally in recent years are not new, but pre-date the Amin years.

At the union level, a similar picture emerges. In 1988 the tobacco union office, whose buildings are shared by the BAT Divisional Leaf Officer and his staff, was very busy. The union has a number of vehicles which are hired to BAT for transport of inputs and tobacco, and used by union staff for touring the district. The production units for pipes and tiles were operating at peak capacity. By contrast, at the cotton union a few workers were desultorily installing machines, although they lacked some of the necessary parts. In addition, the dilapidated stores still bulged with the unginned 1978–86 cotton crop, and the compound was dotted with the relics of a variety of vehicles and machines half-buried in grass and bushes. Several key staff were absent and up-to-date records difficult to find. It was not clear how many societies were still members of the union or even in existence.

Another measure of the level of co-operative activity is the number of meetings held in the year. The mean number of both annual general and management committee meetings are three

185

times and more higher for the tobacco societies.

The successful symbiotic relationship between the company and the co-operatives is the main reason why the growers and the tobacco societies are active and productive in contrast to other co-operatives in Arua, which are virtually defunct: notably the cotton ones, many of whose members go to the extent of actively refusing to grow the latter crop. Another feature of this vertical integration is the involvement of the co-operatives in marketing cigarettes. The union is the district distributor. This gives the growers a substantial interest and share in the profits from the end product of their toil.

The use of co-operatives, and the way in which they are integrated into the organisation of the industry as a whole, has over time introduced new concepts of co-operation and management. It has also helped to lay a basis for political action, allowing local communities to bargain with higher authorities. Other studies have shown that a high proportion of early political leadership was drawn from the ranks of the co-operative movement.

The Economic Impact on Farmers

The introduction of tobacco to West Nile had the immediate effect of providing a good source of income to farmers which had not been available up to that time. It enriched growers who, by the 1960s, were measurably better-off than non-growers.[21] It tended to attract the more educated with larger resources of land and family labour. Returning migrant workers often invested earnings from jobs in the south in tobacco growing. Money from tobacco was in turn invested in other enterprises. According to Hutton (1968),[22] such investments included farming, trading, housing, livestock and other material goods. Thirteen per cent banked their money. In 1988, farmers reported having used income from tobacco for the purposes listed in Table 9.6.

Tobacco curing is carried out co-operatively by groups of four people. The owner, the barn-leader, was and still is usually pre-eminent in production. While it was not possible to break down the income data into categories of barn-leader and others, preliminary interviews suggested that the barn owner markets much larger quantities of tobacco than other barn members. Such barn-leaders were shown in the 1960s to have a higher status in many respects

21. Hutton, *Unemployment and Labour Migration*, chapter V.
22. Ibid., pp. 221–6.

Table 9.6 Investment of farm earnings in other enterprises by growers prior to 1988

Purpose	Per cent
Livestock	74
House	33
Wood lot	23
Transport/car	6
Shop	5
Land	4
Dairy farm	2

than other growers. They had more wives than other growers and even more so than other farmers; 41 per cent had two or more compared with 31 per cent for non-growers.[23] In 1988 34 per cent of all growers still had more than one wife compared with only 16 per cent of non-growers. The desire for more wives was always expressed in terms of cultural values rather than the need for labour, although many informants said that having a large family was necessary for successful growers.

The income derived from tobacco has continued to be important to the livelihood of a large number of farmers and their families. Tobacco is at present the only reliable cash-crop. Delays in payment for both coffee and cotton and problems with the supply of inputs continue to discourage farmers from cultivating them. In 1988 cotton growers in Arua reported that they had only just been paid for the 1984 and 1985 cotton crops. This is particularly serious in the situation of hyperinflation which has existed in Uganda. Tobacco growers, on the other hand, are paid within one month of delivering their produce.

Many food-crops are also marketed, but farmers stated that the market for them is not organised and is small and irregular. Farmers are usually only able to sell about 20 kg at a time, and sometimes they can return from the market with their produce unbought.

It was early on suggested that the higher incomes of tobacco growers would contribute to the emergence of a more stratified class society. However, over time this comparative advantage has not been sustained. Their relative affluence has been eroded as a result of the economic disintegration of the Amin era and the growth of alternative means of earning money. Between 1972 and 1984 the local value of tobacco deteriorated in relation to manufac-

23. Ibid., p. 179.

tured commodities, such as blankets, tools and sugar, because of the overvaluation of the Uganda shilling. There was also a general neglect of the industry by government and the National Tobacco Company.

As Table 9.7 shows, today few growers are materially better off in most respects than non-growers inspite of the fact that they probably earn much more money from farming, since they also grow and sell larger amounts of all other crops, with the exception of sweet potatoes. There is also no evidence today that tobacco growers, or in fact the better-off, form a separate class. Rather, kinship linkages continue to pervade society in a manner that militates against this form of stratification. People still look on their kin of whatever economic status as their primary group of reference, identification and loyalty. Table 9.7 gives the past and present situation of growers and non-growers in relation to selected wealth indicators. It demonstrates both the extent to which total wealth was destroyed and the rate of recovery, as well as differentials between the two groups over time.

The size of earnings in 1987 in the surveyed societies is shown in Table 9.8.

The relatively poor current comparative situation of growers is undoubtedly largely due to the decay of the industry during the 1970s coupled with improved alternative opportunities in commerce and employment, and to the subsequent losses experienced after the War of Liberation; as a result of which returnee farmers' first concern has been to re-establish homes and farms rather than buy other items of conspicuous consumption. However, it may also be that the introduction of a high-value cash-crop involves learning how to manage cash incomes. Tobacco earnings come in a lump sum at one time at the end of the year, and so far few farmers seem able to plan the use of this money. They tend to spend the cash as soon as they get it, and it was remarked that much drinking takes place next to the societies on the days the farmers get paid. Although this may serve to recirculate the money since growers' wives are the major brewers, it nevertheless reduces the total available for other things.

This situation is exacerbated by the present inadequacies of banking services. At one time mobile banks operated out of district headquarters; however, these were discontinued due to insecurity. Up until 1988, therefore, the only bank for the whole of Arua was one branch of the Uganda Commercial Bank in Arua town. As part of the present economic policies of the Uganda government, branches have now been opened in many sub-county headquarters, but in 1989 these were not fully functional and farmers were reluctant to trust them.

Table 9.7 Measurement of economic status by ownership of selected wealth indicators

Item	1978[a]		Percentage owning 1988	
	Growers	Non-growers	Growers	Non-growers
Cattle 1–10	55	45	45	42
11–20	14	8	11	5
Goats 1–10	58	57	75	71
11–20	21	19	9	9
Brick house	8	2	12	7
Bicycles	60	47	52	28
Radios	29	19	14	12
Beds	53	45	26	16
Mattresses	56	43	18	17
Blankets	68	61	40	33
Oil lamps	43	28	12	8

a. Many items reported looted or destroyed in 1980/1, including most houses destroyed. Thus possessions recorded for 1988 were mostly acquired during the 1980s.

Table 9.8 Earnings of growers in survey societies, 1987

Society	0–10		10–20		20–30		'000s Ugandan shillings 30+	
	No.	%	No.	%	No.	%	No.	%
Otrevu	146	45	109	34	58	18	11	5
Ayivu	115	43	83	31	42	15	30	11
Wandi	155	33	170	37	77	17	62	13
Oluffe	247	43	228	40	66	11	30	5

Note: Due to the high rate of inflation officially calculated in 1988 to be 150 per cent per year, but probably nearer 400 per cent on company estimates, these amounts appeared small compared to expected 1988 earnings since the price per kg of leaf had been increased to nearly three times that for 1987.

The Impact on Farming Practices

The cultivation of tobacco as with other cash-crops has introduced several concepts new to African society, many of which have still not been fully absorbed. Tobacco is a crop which requires high standards of cultivation and the adoption of new techniques. It has to be sown in seedbeds using fertilisers during the dry season so that much watering is also needed. It has to be transplanted into rows in fields which have also had an application of fertiliser. The leaves have to be picked at a certain stage of maturity, not all at once. It must be kept clean-weeded. Tobacco should be the first crop to be planted on land which has been fallow. Any cash-crop also has to be managed in conjunction with the production of food-crops, since the scale on which it is grown by peasant farmers does not allow for specialisation. These requirements have had effects both on cropping systems and on labour organisation.

Traditionally, fields were reserved for different crops by location. As was described by Middleton: 'Each homestead, occupied by a single wife, has three types of fields: the *amvu akua* (field at home), the *amvu amve* (fields outside), and the *yimile* (riverine or irrigated fields).'[24] Today, largely as the result of population pressures, few people still maintain plots of different categories, while the general size of land holding is very small as can be seen from Table 9.9

The population of Arua has probably more than tripled since 1936. Available statistics show that for Arua and Moyo it increased from 297,000 in 1948 to 855,000 in 1988.[25] Arua forms roughly two-thirds of this total.

One of the major advantages of tobacco to growers is that the return per hectare is greater than for nearly all other cash-crops grown in Uganda, while proportionally it utilises a very much smaller land area. Nevertheless it was still claimed that only those with larger acreages could be growers. Of those who had never grown tobacco 59 per cent gave land scarcity as the reason. Table 9.9 illustrates that not only do growers on average have bigger acreages but also a larger area under cultivation. Twenty-six per cent of growers own more than 2 ha compared with 11 per cent of non-growers, while 22 per cent compared to 11 per cent were cultivating more than 1.3 ha.

24. J. Middleton, 'The Lugbara of Uganda', p. 11.
25. Uganda Government, Population Census 1948, 1959 and 1969; Background to the Budget, 1988/9.

190

Table 9.9 Size of land holding 1988 and amount cultivated

Hectares	Growers		Per cent Non-growers	
	Total	Cultivated	Total	Cultivated
Less than 0.4	8	17	8	25
0.4–1.2	34	60	49	64
1.3–2.0	32	17	32	9
2.1–4.0	20	4	9	2
4 and more	6	1	2	0

Table 9.10 Crop grown prior to tobacco in 1988

Crop	Per cent of growers
Nothing	50
Simsim	7
Tobacco	1
Maize/millet	4
Beans/groundnuts	3
Cassava	22
Other	11

One result of the introduction of tobacco is that the traditional rotations have been replaced to favour its cultivation. At least half of tobacco growers plant tobacco as the first crop. This is shown in Table 9.10. However, in spite of the long period over which tobacco has been grown, and while recommendations for its cultivation are largely observed, there has been somewhat variable adoption of similar practices for other crops. Ninety-seven per cent use fertiliser for tobacco, but only 12 per cent use it for any other crop. (The latter may, however, obtain some residual benefit from the application to tobacco.) It was also evident that, probably due to cost, the amount of application on tobacco was below that recommended. Most primary societies visited reported large stocks of unsold fertiliser. One other reason for this may be that while fertiliser improves the yield of tobacco from the field, ultimate returns are more dependent on the quality of cured leaf, which is a function of proper control of the curing process.

Other agrochemicals are also mainly used for tobacco; nevertheless some farmers use them on other crops, as is evidenced by the number of non-growers who apply them. Acceptance of innovation

Table 9.11 Use of agrochemicals and improved seeds
by grower category

Product	Growers	Per cent non-growers
Fertilisers	97	7
Fungicides	19	3
Pesticides	26	8
Improved seeds	66[a]	25

a. Not including tobacco.

is also demonstrated by the number of farmers using improved
seed for such crops as maize or vegetables (see Table 9.11), and the
proportion who used improved agronomic technologies in the
past. As may be seen, there is a very marked difference in the
percentage of growers using improved seed over non-growers.
Fifty-five per cent of growers also said that they had used tractors
in the past as against only 26 per cent of non-growers. Today there
are almost no tractors in the district, and growers do not have
access to them, although there are proposals to assist the tobacco
union and societies to obtain new tractors to hire to members.
Ox cultivation was not an option that any grower used in the
past, although its introduction is now also being considered in
view of the acute shortages of fuel to which the district is presently
subject.

Tobacco requires extensive amounts of fuelwood for curing.
Thus the growing of tobacco has to be accompanied by agro-
forestry, or silviculture, for this purpose. As early as 1954 the need
for such plantations was recognised and growers' primary co-
operative societies were later established for the purpose of planting
and managing forest plantations. The demand for woodfuel for
curing focused attention on the need to have self-sustaining supplies through forest plantations long before environmental issues
were of concern in Uganda. They continue to be a vital element of
tobacco culture in Arua. Each co-operative society has one or more
plantations of its own.

Theoretically, 7,225 ha are owned by the 22 tobacco societies.
However, there has been substantial deforestation due to fires,
uncontrolled felling for other purposes and lack of replanting. The
actual hectarage today is estimated to be about 3,846 ha. Some
societies' plantations are now virtually extinct, forcing members to
seek fuel quite far afield and placing increased pressures on the

remaining societies' forests, mainly those in Maracha. It was also evident that even the gapping being carried out in the plantations was not only inadequate but poorly maintained. After planting, a tree survives by natural chance rather than any sustained care. Even though all recognise that reserves are being rapidly depleted and that some society plantations are now exhausted, growers are unprepared to do much work in their society plantations, and the low level of wages offered for such work makes it difficult to get anyone else. More recently farmers have been encouraged to grow their own trees. So far only a very few have responded. Before the exodus in 1980–1 there were 200 growers with 264 ha of private plots. Today the same growers have only 70.5 ha.

Tobacco growing requires the attainment of very specific quality standards. It was at first thought the semi-literate peasant was incapable of such sophistication. However the quality of Arua leaf is recognised by the experts to be high. This is encouraged by paying premium prices for higher grades. At one time there were 36 quality grades, today this has been reduced to 12. While tobacco is graded by the company buyer in the presence and with the approval of society officials, the farmers themselves do the initial sorting and are fully conversant with the criteria by which standards are assessed.

The Impact on Farm Resource Management

Tobacco production is labour–intensive. The various operations are also spread over the whole year and are not confined to a limited number of months as with other crops. It is claimed by growers and others that only those with large families can manage the cultivation of the crop. There are two levels at which co–operation is required in tobacco leaf production in Arua. The first, within the nuclear family, is that needed for labour mobilisation for cultivation. The second, for curing, involves several, sometimes unrelated, families sharing the use of a capital asset.

In mobilising labour for the cultivation of tobacco, farmers utilise all members of the family. It is therefore not only the women, as is discussed further below, who take on additional responsibilities. Children are also needed for all stages of the cultivation and curing process and particularly for such tasks as weeding and stringing. Tobacco production may therefore be a factor, though unstated, in fewer of the children of growers being in school, as well as their late entry into primary school. It may also result in children being dependent for longer than in non-grower households. The labour needed for tobacco furthermore reduces its availability for other crops, and was a main reason given by women for not growing the crop themselves. Only two women growers were identified in the sample and both had temporarily abandoned the crop to concentrate on the cultivation of food.

In Arua, unlike other flue-cured tobacco areas in Uganda, four people are required to share a barn. The barn is presently considered to be the property of one of these four on whose land it is situated, although initially it appears that there was group ownership, and payments for barn rehabilitation and maintenance are deducted from all who use it. In many instances the other members are relatives, though not always immediate ones, but in others they are friends or neighbours. In 1970, Hutton showed that there were non-kin in only 5 per cent of barns.[26] In 1988, in 83 per cent of barns all members were related to each other, while in 17 per cent some members were friends or neighbours. This represents a measurable shift to other criteria for co-operation than kinship.

This type of co-operation does not build on any local tradition, and there are sometimes conflicts between members. Access to a barn is fundamental to becoming a grower, and barn owners largely control who else also grows tobacco. Five per cent of non-growers who had at one time grown tobacco and 31 per cent of those who had never grown the crop said that this was due to lack of a barn. This, therefore, puts the barn leader in a dominant position within the family and the community, particularly since no new barns have been built since 1972. There is no information on how the original barn-owners came to be chosen, although it may be assumed that they included known more reliable and larger growers who had land available for the site. Most of the present owners are the original ones. Extremely few barns have been inherited by eldest sons. There are at present, 2,800 barns in operation out of a possible 3,692 which had been built up to 1979.

The Social Impact

The social impact of the tobacco industry is not clear-cut, and it is difficult to separate cause and effect. As with other innovations it can be assumed that ideas, attitudes and human relationships may be changed. The dynamic of these changes is not, however,

26. Hutton, *Unemployment and Labour Migration*, p. 184.

straightforward. For instance, it is generally believed that more educated farmers are more likely to adopt innovations successfully. In the case of tobacco in Arua it is suggested that the reverse has often been the case.[27] An early attempt to involve the educated élite in flue-cured tobacco growing failed because 'their attention was divided between trade or local politics or teaching and tobacco cultivation'; and it was, therefore, 'the ordinary illiterate peasant cultivator' who it was thought 'would be unable to cope with somewhat sophisticated techniques of curing' who eventually was recruited. Although a high proportion were illiterate, nevertheless the figures in Table 9.12 demonstrate the educational lead growers had over non-growers shortly after flue-curing became established.

In the 1960s it appears that the more educated joined tobacco growing, and that growers were also more committed to education for their children. Over time, however, they have largely lost this advantage. Although the overall educational level has substantially improved, it is the non-growers who appear to have taken more advantage of the increased accessibility of schools. An attempt was made in the case of farmers' wives to assess the influence of age on this variable. Nevertheless the disparity still holds as is shown in Table 9.13.

Another negative effect seems to be on ideas and attitudes. It was found in 1988 that growers had a smaller proportion of their school-age children in school than non-growers although this may also be because they have larger families. Growers were maintaining three children on average in school versus two for non-growers. Table 9.14 compares the situation in 1988 with that in 1968. In addition, while both growers and non-growers accorded low priority to girls' education in 1968, this had been reversed for non-growers in 1988, but still persists for growers although to a much lesser degree.

Wives of growers were more conservative in outlook, having fewer modern aspirations for their daughters than the wives of non-growers. While more growers' wives had wanted to go further in school than those of non-growers, only 57 per cent compared to 80 per cent would like their daughters to have a career. In addition, only 17 per cent as compared with 49 per cent thought that girls were equal to boys in educational and job potential. This difference still held when controlling for age and education.

The relative conservatism of growers and their wives is difficult

27. Mackenzie. *Present Location and Past Diffusion of the Flue-cured Tobacco Industry*, pp. 180–2.

Table 9.12 Education level of farmers by grower category, 1968 and 1988

	1968		Per cent 1988	
	Growers	Non-growers	Growers	Non-growers
None	47	72	15	17
Primary	49	27	66	62
Post-primary	4	1	19	21

Table 9.13 Education of farmer's wives by grower category

	29 and under		Per cent 30 and over	
	Grower	Non-grower	Grower	Non-grower
None	41	39	65	43
P1–3	30	10	24	21
P4+	29	51	11	36

Table 9.14 Percentage of school age children in school by grower category: 1968 and 1988

	In school 1988		1968[a]	
			Sons	Daughters
		Barn leaders	95	60
Growers	75	Other growers	82	45
Non-growers	85		58	31

a. Barn leaders were listed separately from other growers, as were sons and daughters by Hutton (1968: 180–2).

to account for. However, tobacco growing may be contrasted to migrant labour, the other main source of income in the past. It was shown that labour migration was associated with the breakdown of many cultural traditions. It may, therefore, be argued that local alternatives to earn money may have strengthened, rather than weakened, customary behaviour and attitudes.

The Impact on Women

Another potential area of change is that of the relative roles, status and responsibilities of men and women. In the case of tobacco in Arua, men who previously had little to do with most agricultural procedures have had to become more fully involved in cultivation. Their wives, who had always been almost solely responsible for the production of food-crops, have had to extend their working day to include participation in growing tobacco. This has implications beyond the actual labour involved. In general, male dominance has been reinforced.

Women were traditionally responsible for the supply of food-crops and this continues to be the case, while they are now also expected to contribute to the cultivation of their husbands' tobacco. Some tasks such as watering and porterage are almost solely the province of women and they are also to a larger or lesser degree involved in almost all other activities (see Table 9.15).

Growers' wives consider that tobacco has added to their workload, while not improving their social and economic status. There are almost no women growers, partly because it is difficult for women to combine cultivation on their own account with helping to cultivate their husbands' plots and other farm work, but it may also be due to the fact that women still do not own land and may have particular difficulty getting access to a barn. The latter is entirely dependent on the goodwill of husbands.

Growers' wives made several general statements about the consequences for them of growing tobacco. Very few expressed positive feelings about the crop. Only 29 per cent felt that tobacco families were better off than others, although they admitted that tobacco brought in more money for family needs and nearly half reported that their husbands had purchased things for themselves and their children over the past year.

Seventy-five per cent of growers' wives agreed that tobacco makes more money than other crops, and 69 per cent said that it paid the school fees, 79 per cent that it enabled them to hire extra farm labour, and 80 per cent that it paid for any major expenses, while 79 per cent thought that they were helped to buy more of everything. Even more cogently, 47 per cent of growers' wives said that their husbands had recently purchased clothes and other items for the family as against 20 per cent of non-growers. Nevertheless, there is no very large, visible difference in standards of living between the homes of most growers and those of neighbouring non-growers, and on one measure of economic status, that

Table 9.15 Level of participation of women in tobacco cultivation and curing in Arua

| | | Per cent | |
		Always	Sometimes
Seedbed work		7	1
Cultivation:	Digging	21	17
	Harrowing	74	15
	Transplanting	91	1
	Watering	90	4
	Weeding	21	18
	Harvesting	75	7
Curing:	Transport to barn	93	2
	Stringing	60	16
	Loading barn	29	18
	Cutting firewood	6	5
	Tending furnace	4	5
Marketing:	Bundling	54	13
	Transport to society	86	4

of ownership of personal possessions, growers' wives appeared in general less well off than those of non-growers.

Sixty-six per cent of women disliked growing tobacco because men control most of the income from it. Fifty-nine per cent of husbands decided how to spend tobacco money as compared to 22 per cent for other crops, and 30 per cent compared to 18 per cent spent the money alone. In contrast, 40 per cent of growers' wives kept the money from selling food crops, but none from the sale of tobacco.

Sixty-five per cent of wives did not like growing tobacco because of the amount of time it takes to cultivate. More than 85 per cent stated that it gave them less time for all other activities. Their responses possibly tend to exaggerate the negative impact of the labour involved in its cultivation since, as may be seen in Table 9.16, a higher percentage of wives of growers were actually involved in other income-generating activities, particularly brewing and the sale of processed food, even though they felt they had less time for 'other farm work' and 'other activities'. It must, however, be pointed out that these activities are not full-time and no information was obtained on the amount of time given to any one of them. Brewing, for instance, may be undertaken very irregularly, while tailoring is more usually a daily enterprise.

No information was obtained on how much money was made

Table 9.16 Sources of wives' independent income by age and grower category

	<30		Percentage 30+	
	Growers	Non-growers	Growers	Non-growers
None	4	4	9	4
Trading	7	9	4	18
Brewing	73	48	76	36
Tailoring	2	2	0	14
Sewing/embroidery	11	7	5	14
Mats/baskets	9	6	15	9
Sell cooked food	7	7	6	18
Casual labour	9	17	14	36
Sell crops	32	24	40	41
Sell process foods	39	11	33	9

by women from any source; nevertheless women's own earnings are important to the family. Women say they use personal income for the home generally, and spend little on themselves. Items for family use are mainly such things as soap and salt, and educational materials for children. There was little difference in the responses of growers' and non-growers' wives on this issue, although slightly more of the latter said they spent their money on a variety of other things as well as for the home generally.

One other factor may serve to explain the lack of any positive advantage for growers' wives in respect of wealth indicators. It has been shown in other parts of Uganda[28] that brewing is coincident with access to larger land-holdings, and related to the need to provide beer and food for extra labourers. This may also be the case in Arua, since a primary investment of earnings is in agriculture and such brewing may not have generated income primarily for personal use.

Even allowing for all ameliorating considerations it does not appear that women have improved their status as a result of involvement in tobacco cultivation. They have more work for little direct personal reward, although the family as a whole may be materially better off. The introduction of cash-crops often has a negative impact on women and this seems to be the case with tobacco too.

28. J. Vincent *Teso in Transformation: The political economy of peasant and class in Eastern Africa*, University of California Press, 1982.

Conclusions

In reviewing the effects of the introduction of tobacco over time, it is evident that these have been far-reaching and that, although it is no longer a new crop, changes in the organisation of the industry and modifications to agronomy and curing are continuing. These cumulative changes over time have had, and are continuing to have, important implications for rural change in the area.

Cash-crops were first grown in Uganda as a means of justifying the expense of administration. In Arua their introduction was claimed to be an attempt to offset the flow of young men out of the district, which was having a destabilising effect on rural communities. While other crops including cotton and sunflower were grown prior to the introduction of tobacco, the latter represented the first real opportunity for farmers to make a reasonable income locally. Migration also brought money to Arua, but it is suggested that local income-generation made a more lasting contribution to the development of the economy of the region. It assured the integration of Arua into the mainstream of the national and indeed the world economy, and also effected a real improvement in standards of living, at least in the form of material wealth. It not only served to make accessible a wider range of goods to growers themselves, but made a positive input into the lives of other farmers.

Tobacco was also instrumental in introducing farmers to scientific agriculture. At the same time, it involved them in modern management systems and new avenues to leadership and power through the co-operatives. Incidental to its growth were the development of roads and transportation, and of forestry projects. Tobacco was one of the most important influences on Arua in the decade before the Second World War, and continued to be a vital factor in regional development up to 1972. Since 1984 it is again a key factor in the regeneration of the economy of the area.

Appendix

Table 9.17 Official exchange rate, 1981–9

(New Uganda shilling per US dollar)[a]

	1981	1982 Window 1	1982 Window 2	1983 Window 1	1983 Window 2	1984[b] Window 1 (auction)	1984[b] Window 2	1985 Auction	1986 Fixed priority	1986 Market	1987	1988	1989 Ordinary	1989 SIP[c]
Jan	0.076	0.860		1.070	2.350	2.41	2.99	5.39	14.79		14.00	60.00	165.00	
Feb	0.078	0.860		1.120	2.300	2.53	3.03	5.57	14.80		14.00	60.00	165.00	
Mar	0.079	0.858		1.166	2.360	2.74	3.20	5.59	14.80		14.00	60.00	200.00	
Apr	0.080	0.859		1.197	2.680	2.92	3.18	5.93	14.80		14.00	60.00	200.00	
May	0.081	0.874		1.283	2.755	2.92	3.26	6.00	14.00	50.00	14.00	60.00	200.00	
Jun	0.768	0.942		1.388	2.900	3.07		6.00	14.00	50.00	60.00	60.00	200.00	400.00
Jul	0.794	0.987		1.551	2.937	3.38		6.00	14.00	50.00	60.00	150.00	200.00	400.00
Aug	0.817	0.992	3.000	1.665	2.936	3.74		6.00	14.00		60.00	150.00	200.00	400.00
Sep	0.805	0.991	3.000	1.763	2.710	4.00		6.00	14.00		60.00	150.00	200.00	400.00
Oct	0.791	0.997	2.000	1.856	2.955	4.46		6.80	14.00		60.00	150.00	340.00	
Nov	0.783	1.017	2.520	2.067	3.260	5.46		8.70	14.00		60.00	150.00	370.00	
Dec	0.856	1.048	2.400	2.338	3.025	5.52		12.75	14.00		60.00	165.00	370.00	

Source: Research Department, Bank of Uganda.
Notes: a. The new shilling was introduced in July 1987.
  b. Windows unified in June 1984.
  c. Special Import Programme 2.

References

Bohanan, P. and Dalton, G., *Markets in Africa*, Northwestern University Press, 1962

Dodge, P. C., 'The West Nile Emergency', in P. C. Dodge and D. P. Wiebe (eds), *Crisis in Uganda: The breakdown of health services*, Oxford, 1985

—— and D. P. Wiebe, *Crisis in Uganda: The breakdown of health services*, Pergamon Press, 1985

Hutton, R. C., 'Unemployment and Labour Migration in Uganda', PhD Thesis, University of East Africa, 1968

Mackenzie, M. K., 'Present Location and Past Diffusion of the Flue-cured Tobacco Industry in West Nile District', Occasional Paper No. 21, Department of Geography, University of Makerere, 1971

Middleton, J., 'Social Change among the Lugbara of Uganda', *Civilization*, Brussels, vol. 10, no. 4, 1960

——, 'Trade and Markets among the Lugbara of Uganda', in P. Bohanan and G. Dalton (eds), *Markets in Africa*, Evanston, Illinois, 1962

——, *The Lugbara of Uganda*, Holt, Rinehart & Winston, 1965

Mwaka, V. M., 'Agricultural Marketing Cooperatives in Uganda', Occasional Paper no. 69, Department of Geography, University of Makerere, 1978

Pain, R. D., 'Acholi and Nubians: economic forces and military employment', in D. P. Wiebe and P. C. Dodge (eds), *Beyond Crisis: Development issues in Uganda*, Makerere Institute of Social Research, 1987

Purseglove, J. W., *Tobacco in Uganda*, Uganda Government Printer, Kampala, 1951

Uganda Government, *Background to the Budget*, 1988/9

—— *Population Census*, 1948, 1959, 1969

Vincent, J., *Teso in Transformation: The political economy of peasant and class in Eastern Africa*, University of California Press, 1982

Wiebe, D. P. and P. C. Dodge, *Beyond Crisis: Development issues in Uganda*, Makerere Institute of Social Research, 1987

World Bank, Project Completion Report on the Uganda Smallholder Project, IBRD, 1982

10

Issues of Energy Autarky and Interdependence among Small-scale Commercial Farmers in Zimbabwe

Angela Cheater

Introduction

Practically everyone would agree that the teleology of underdevelopment revolves around low and technically inefficient energy use. More energy must be used more efficiently, in new technologies, if Third World countrysides are to develop. But there are problems of the availability, reliability and cost of 'modern' forms of energy and the equipment required to use such new forms of energy. As examples: in Zimbabwe the past pricing policy for electricity meant that producers (e.g. in aquaculture) immediately under the power lines at Kariba paid much more for their power than did urban domestic consumers hundreds of kilometres distant; in China (and probably elsewhere) the first to be switched out of the national electricity grid when demand peaks are village industries; and in the commercial farming sector of Zimbabwe, the nearest fuel station for petrol, diesel and paraffin is commonly 20–50 km distant. Only a quarter (150 of the total of 620) of Zimbabwe's fuel stations are in the countryside (*Herald*, 14 July 1989, p. 7), an infinitesimal fraction of these serving peasant farming areas. Large-scale farmers, therefore, have safe and approved on-farm storage facilities replenished by tanker deliveries; a few smaller farmers store petroleum products in less safe drums and jerrycans; while for most peasant producers petroleum fuels, as sources of their energy needs, are as irrelevant as electricity.

In order to rectify past inequalities and alter this situation, shortly

I am grateful to Des Gasper, Shanti George, David Hancock and Richard Harlen for comments on an earlier draft of this paper and discussion of the issues raised.

203

after Independence in 1980, the Zimbabwe Energy Accounting Project (ZEAP) attempted, as a type of integrated energy planning, 'the most detailed assessment of Zimbabwe's energy situation undertaken to date'. While not presuming to define 'an authoritive statement of energy policy in Zimbabwe', this study did attempt to identify future energy problems and their possible solutions (Hosier, 1986: 1). It estimated that rural energy consumption accounts for over 60 per cent of the total energy consumed (Hosier et al., 1986: 15), the vast bulk apparently by peasant cooking and housing. However, Hosier et al. (1986: 21; 1988b: 32) admit that, because their quantification was problematic, both draught and human power were omitted from ZEAP's calculations, even though both are crucial to the majority of Zimbabwe's farmers. Yet, in a separate exercise, ZEAP extrapolated values for these energy sources from the LDC Energy Alternative Planning (LEAP) computer program developed for Kenya.[1] For the 1981–2 cropping season, draught power was estimated by this method to total 3.6 million gigajoules (GJ) and human energy supplies 2.5 million GJ, compared to a survey-derived figure of 2.78 GJ for petroleum products consumed by (mainly commercial) agriculture during that season. Were such estimates for draught and human energy to be included in ZEAP's calculations, Zimbabwean peasant agriculture would account, not for the 1 per cent of all energy consumed in agriculture that it is allocated, but 25 per cent (Hosier 1988b: 35). So much for integrated energy planning in theory and in practice!

The ZEAP analysis does indicate that it is important to comprehend 'the flows of energy between and within' the sub-sectors of rural production in Zimbabwe (though it is debatable whether it achieves this); and that more energy must be supplied if the currently underdeveloped areas of the countryside are to develop further (Hosier, 1988b: 14). None the less, given the omissions noted above, the ZEAP analysis has in practice led to a policy orientation for rural energy development which focuses immediately on fuelwood and the development of low-cost cooking devices to replace open fires (ZEAP, 1985; GTZ, 1987), and ultimately foresees rural electrification (Temane et al., 1987).[2]

1. LEAP's 'fuelwood bias' has been criticised as inappropriate to (many parts of) Zimbabwe (GTZ, 1987: 11).

2. On purely technical grounds, this emphasis is misplaced. David Hancock has found, in woodfuel-deficit areas close to Harare, that new local techniques of using open fires may use less fuel than newly developed 'fuel-efficient' stoves such as his own award-winning design (personal communication)! It is difficult, however, to convince Third World governments that the past emphasis on 'the other energy

However, energy provision for rural society is not homogeneous, nor are all rural people involved in identical production and consumption. There are different 'development thresholds' (George, 1988) within a single countryside, and their energy needs for further development are quite different. We therefore need to understand, at a much more differentiated and micro-level, precisely what kinds of energy are used, by whom, for what purposes, at what cost relative to output value, and how different energy sources and needs are traded off against one another – how energy flows around and through the different local systems – before rural energy planning can sensibly be instituted.

This chapter, therefore, examines energy interconnections in one small-scale commercial farming area in Zimbabwe, as these existed in the mid-1970s.[3] These data are important because such farmers' energy and technological requirements are also typical of (potentially very many more) successful peasant producers, both now and in the future. There are fewer than 10,000 small-scale freeholders in Zimbabwe, and their scale of operation is, at a national average farm-size of some 80 hectares and in relation to the majority of farmers elsewhere on the sub-continent, large. However, in areas where land is not a constraint to production the example of such producers is not irrelevant to sustainable agriculture. Nor should we completely overlook the significance of success while focusing attention on those struggling to survive. It is doubtless unnecessary to add that the 1974 figures for cost prices and revenue bear no relationship whatsoever to current reality. However, by taking these data as representative of agricultural success on a limited scale in an area where fuelwood is not an important energy constraint, in our view we may achieve a better, albeit qualitative, understanding of differentiated rural energy demand and its associated technology, than that offered by ZEAF (Hosier, 1986, 1988a, b).

---

crisis' has been misplaced and that new approaches are required. In Zimbabwe, some attention has been paid to reducing women's work by cutting distances to public water supplies, by sinking boreholes; and to the provision of animal draught and especially (state-owned) tractor power in resettlement schemes. However, neither has been part of an overall rural energy policy, but rather specific responses to specific problems defined by different ministries, and both are currently experiencing difficulties on a substantial scale.

Despite differential demand in those centres already electrified, often selected (it would appear) on political as much as economic grounds, no evaluation of the rural electrification programme has yet been undertaken (five years after its commencement), nor has the parastatal Zimbabwe Electricity Supply Authority's master plan for rural electrification been accepted by ZESA's parent ministry (personal communication, Richard Harlen).

3. The data analysed were collected by the author in the mid-1970s.

Components of the 'Energy System' in Msengezi

We have examined the various facets of production, marketing, the distribution of output value and social organisation in Msengezi in a number of earlier publications (Cheater, 1974, 1976, 1978a, b, 1979a, b, 1981a, b, c, 1982, 1983, 1984, 1985, 1987). Suffice it to note here that, by 1974, modern sources of energy were under-developed even after forty years of settlement and successful *commercial* production in this freehold area with its relatively good services and facilities. In 1973 not even the local council offices had electricity, although one farmer used a diesel generator in preference to paraffin or gas to light his house. By the end of 1974, his was among the four farms very recently connected to, or in the process of negotiating connection to, the national electricity grid, and the local council was hoping to take financial advantage of their initiative. There were two stations supplying petroleum fuels: one with underground tanks and manually-operated liftpumps on the western periphery of Msengezi and the other, relying on 2,000-litre raised storage tanks, in the centre of the area. Apart from these local outlets, which occasionally ran out of supplies, the nearest fuel was in Chegutu or Seious, respectively 40 km and 30 km from the centre of Msengezi. Not surprisingly under these circumstances, coupled with the very restricted availability of new farm machinery, only 54 of 338 farms and smallholdings owned tractors (Cheater, 1984: 82); only 34 per cent of the total cultivated area was ploughed by tractor (Cheater, 1981: 368); and even the local council had acquired its first motorised road grader only late in 1962. People and animals remained – and today still remain – the prime sources of energy for a wide range of competing developmental needs. The farmers' main headache also remains how to allocate the available human energy resources and draught animals among these competing needs.

The energy requirements for further development in Msengezi can be grouped into three major categories: domestic energy required for social reproduction; energy required for the development and maintenance of the productive infrastructure (roads, boreholes, reservoirs, irrigation and soil conservation works), both public and private; and energy used in the production of both crops and livestock in a mixed farming system. Each category has its own, separate, technological corollaries, which are reflected, *inter alia*, in different strategies for production and marketing. In 1973–4, the most common productive strategy was extensive cropping for sale to state marketing boards, backed by capital reserves in the form of cattle, with an annual sales offtake of some 20 per cent (Cheater, 1984: 46). A minority of farmers, however, preferred higher risk and higher return 'entrepreneurial' strategies of production and marketing (Cheater, 1984: 44, 50).

In order to give some idea of how and why decisions on energy sources and use differed from farm to farm, I shall examine two case-studies in detail. Farmer *A* was a relatively successful example of the dominant strategy, farming maize, cotton (the main cash-crop in Msengezi) and beef cattle. In contrast, *B* (whose farm lay on the boundary of Msengezi, less than 2 km down the main road from an old-established, mission boarding secondary school) had rejected the 'safe' strategy of field-crop production for sale at guaranteed marketing board prices, in favour of exploiting a limited and localised market for perishable commodities. His two concessions to 'safe' marketing were the production of pigs for sale to Colcom (a large co-operative company) and fattened cattle sold to the Cold Storage Commission. His other two lines were much more risky: fresh vegetables (admittedly, with the exception of rape, not highly perishable) and fresh milk supplied on contract to the mission's two schools. (There was also a cheaper local market for what went sour before sale, but no labour was put into souring it under controlled conditions.) These four production lines each contributed roughly equal sums to *B*'s total income.

Notwithstanding their differing production strategies, *A* and *B* shared certain characteristics. Both were monogamists in their sixties, and second-generation Christians. *A* was the youngest of three sons of a Tswana immigrant transporter (one of the 'black pioneers'), and was born and raised on a mission station. *B* was a younger son of a Zezuru peasant farmer. Their formal education in neither case exceeded primary schooling, but both had concurrently acquired practical skills, in agriculture and carpentry respectively. Both had had a formal agricultural training (*A* a primary school agricultural certificate and *B* a Master Farmer Certificate, acquired fifteen years after settling on his farm). Both had spent approximately twenty years in formal employment. *A* as a constable and later sergeant in the colonial police force, doubling as court interpreter, and *B* as a mission-employed carpenter. Both had settled on their farms shortly after the end of the Second World War, although *A* had bought his much earlier in 1936. Neither had much capital on settlement, and what they had was mainly in the form of livestock. Nor did they have much personal experience of farming, having acquired their farms mainly on the recommendations of their white employers. Both had relied on semi-clientelistic relationships with white neighbours in building up their farming enterprises: *A* had grown groundnuts to supply a local market for workers' food rations and stockfeed on farms owned by whites, and *B*'s long-standing dairy and vegetable specialisations rested on his personal links to the mission going back to 1939. In the past both farmers had also used their farming operations primarily to finance their children's education (12 of their total of 16 children had completed at least four years of secondary schooling, and 9 held tertiary qualifications). Both had deferred investment in their farms until these educational requirements had been met. As *A* put the point: 'All we did was take out, for education. . . . I believe in educating my children, because if you educate them, they don't become parasites on you later.' Indeed, some of the recent investment, especially in housing on their farms, had been funded by their well-educated daughters, as well as their sons.

By 1974 both farms were classified as 'semi-capitalised' and were therefore among the most successful third of all Msengezi farms (Cheater, 1984: 84). However, neither farmer had opted to mechanise his major productive operations, instead remaining dependent on human energy and draught animals to an extent which, at first glance, seems surprising. Table 10.1 indicates the distribution of labour-time on these two farms among the different use-categories. During the monitoring period, farm *A*'s workforce comprised two adult men (the farm-owner aged 69 and his sister's son aged 23), two grandsons aged 10 and 13, and three adult women (his wife aged 65, a daughter of 28 and a daughter-in-law of 24 on a week's visit to the farm). Farm *B* counted among its adult male workers the owner (aged 62), a hired worker (aged 36), a son (aged 25) on vacation from university and a grandson of 18, plus the farmer's wife (aged 59) and daughter (aged 20). Neither farmer hired casual assistance during the monitoring period, but both had done so shortly before and would do so again on a number of occasions during the growing season. The absolute figures for total labour-hours per day in different uses are, therefore, less reliable than their proportions of the total work-time.

Domestic Energy

Because their labour was so critical in field production, the development of the domestic infrastructure had generally freed women in Msengezi from many routine, time-consuming, home-provisioning tasks, which remain the responsibility of women elsewhere in Zimbabwe and the Third World (Cheater, 1981). Few

Table 10.1 Average daily total labour hours (Jan.–Feb. 1974) on two Msengezi farms

Labour category	Farm A (hours)	%	Female % of total	Farm B (hours)	%	Female % of total
Domestic (including output processing)	9.5	21.9	100.0	19.4	42.9	86.8
Infrastructural	1.3	2.9	0	0	0	0
Off-farm[a]	3.5	8.0	14.3	9.3[b]	20.5	18.8
Field–cropping	20.3	46.3	6.8	9.0	19.9	4.8
Livestock (including milking)	9.1	20.9	15.1	7.6	16.7	0

a. Includes marketing, meetings, social visiting and church activities.
b. Excludes four days spent outside Msengezi by the farm–owner.

Msengezi women had to walk even a couple of hundred metres for water, domestic supplies having been secured to the homestead itself by well or borehole on 65 per cent of all farms (including the 5 per cent with diesel-powered pumping facilities), the remainder being situated within easy access of river water or, in 10 per cent of cases, using a 2 × 200-litre ox-drawn water cart to bring supplies in. During the dry winters, or when new lands were being stumped of trees for cropping use, fuelwood was cut from on-farm reserves by young men and stocked at the homesteads; and was used, in wood/coal ranges or 'Dutch ovens', more sparingly than is characteristic of open traditional kitchen fires (Hosier, 1988b: 88, 219–20).[4] Whole grain retained for domestic consumption and stockfeed was taken (in large sacks, on bicycles or in wheelbarrows, usually by young men) to one of Msengezi's thirteen privately-owned, diesel-powered mills for grinding. During the busy season in large, especially polygamous, households, even cooking, house-cleaning, laundry and child-care were organised by older women to free their reproductive juniors for field-work (Cheater, 1981: 356). While many women certainly carried the double burden of both field and domestic labour, the latter had deliberately been rationalised and made more efficient in many ways to reduce its competition with the demand for human labour in the fields.

4. David Hancock notes that Dover ranges are fuel-inefficient, but in Msengezi (to judge from my own inexpert observations) they used less fuel overall than did open fires. They were also preferred for their modernity and relative cleanliness, as well as their provision of winter heating and year-round hot water as 'by-products' of their use for cooking.

However, domestic labour remained important. On monogamists' farms, there was less scope for scale economies in food preparation or child-care. Labour-time allocations on farms *A* and *B* (see Table 10.1) show that, in the busiest agricultural season, domestic work accounted for over 20 per cent and 40 per cent respectively of all work done; and that it was done overwhelmingly (100 and 86.8 per cent, respectively) by women. (The variation between the two farms in the number of hours of housework (see Table 10.1) arose partly because farm *B* required an average of two hours a day to be spent in washing, weighing and bundling vegetables and cooking green maize for sale, but mainly because of a more fastidious laundry and house-cleaning regime.) While Msengezi women had shed their 'traditional' tasks of hauling water and fetching and chopping firewood (now done by men), they had become more housebound in respect of 'their' work as they aged and were freed of field labour. Attitudinally, too, there were changes, with housework being seen as more appropriate than field-work for 'educated' women of all ages, and some men sharing household chores. On unscheduled visits, I occasionally found a farm-owner polishing cement floors on his knees, while many adolescent sons were proficient in all aspects of housework. 'Clean', electrified housework was even regarded ideologically as appropriate for elderly men, who would have no objection to making their own tea and meals, so they said, if it weren't for the trouble of lighting the Dover stove!

However, the few clearly expressed preferences for electricity were less for its ability to reduce labour on housework than for productive and entertainment (television) purposes. Large-scale poultry production, in particular, was eased by electric power for temperature control. For domestic lighting, paraffin-fuelled pressure lamps were standard among Msengezi farmers, and very much cheaper than the installation charges for electricity.[5] Electric power

5. In 1968, one Msengezi farmer spent a total of £307 on installing a diesel generator. In 1969 Zimbabwe went metric at $2 = £1 (both Rhodesian). In 1974, he spent over R$1,000 on having electricity installed, having split the cost of extending the transmission line with his neighbour, and thereafter paid over R$30 per month on consumption charges. The 1973 quotation to extend a 24 kv powerline from the local boarding school to the Council Offices, a distance of some 6 km, was well over R$4,000, with a minimum monthly consumption charge (for nearly 2,500 units) of R$40. These charges were based on rural consumers paying the full cost of their connection to the national grid, a pricing policy which has subsequently been changed so that urban consumers today subsidise their rural counterparts to a limited extent. But the nominal costs of connection have probably not changed much, given substantial recent inflation. (In 1980 Z$ superseded R$.)

210

Table 10.2 Case-studies: the rationalisation of domestic energy
consumption

Energy requirement	Energy sources	
Fuelwood sources	Indigenous on-farm	Indigenous on-farm + eucalyptus plantation
Cooking/ heating water	Dover range + 200-litre wood-fired bathing system	Dover range
Obtaining water	53ft lined homestead well (manually operated)	32ft lined well (at dairy) + borehole planned
Transportation	1 scotchcart (animal-drawn)	1 scotchcart (animal-drawn)
On-farmoutput processing	Soured milk (amasi): domestic consumption + occasional sale: human energy	1. Washing weighing and bundling, vegetables sold; human energy; 2. cooking green maize for sale: human energy plus fuelwood.

was only for the wealthy, seeking to reduce taxation levels on their
off-farm income or to enjoy the fruits of accumulated capital while
they were still alive. It was not an option dreamed of by the
majority of land-owners.

Infrastructure Maintenance

In order to produce and market their products, commercial farmers
in Msengezi relied on both public and private infrastructure, the
maintenance of which consumed substantial amounts of human
energy, often during the busiest cropping seasons. A flooded
farm-road drain or a burst contour ridge, resulting from heavy
rains, had to be repaired immediately to prevent yet more costly
damage to yields as well as labour-time. Farmers *A* and *B*, like one
in four farmers, both owned ox-drawn dam-scoops, used primarily
for building and repairing contour ridges in a technique invented in
1967 by another Msengezi farmer (Cheater, 1984: 38), and recently
adapted to repairing gullies by similar local inventiveness.

It is the local council's responsibility to construct and maintain
the public infrastructure, for which it receives government grants
of various kinds as well as rates from the land- and business-owners

in the area. However, the parastatal Natural Resources Board also has an ecological policing interest in rural infrastructure, and the committee members of the four Intensive Conservation Areas into which Msengezi is divided had to allot a significant portion of their time to ICA meetings and inspection duties on behalf of the NRB. Committee meetings were normally held monthly, and members were responsible for physically inspecting and reporting on a wide range of ecological issues affecting both public and private land: gullying, other forms of soil erosion, noxious weed growth, 'vermin' and the control of animal pests, the integrity of earth-walled dams (often threatened by children digging for fishing worms!) and contour ridges, etc. The flavour of such activities is reflected in the obligatory 'ward reports' section of ICA Committee minutes: 'Owner of farm no. 47 is said to have tried to intercept the gully which is grossly making a way through his farm, but he still emphasises that something must be done' (Msengezi East ICA, 8 May 1969). In addition to meetings and monitoring ecological concerns in both public and private domains, the ICA committee members were also involved in organising field days and exhibits for the local and Harare shows,[6] and in lengthy external visits. The owner of farm *B*, for example, was away on an ICA visit to Sanyati for four of the seven days during which I monitored labour-time on his farm. His off-farm travels were facilitated by his investment in a small, fuel-efficient motorcycle, fed from jerrycan stocks, which cut to a minimum his own time spent locally off-farm.

Infrastructural and ecological issues, then, consumed a fair amount of the time and managerial energy of male land-owners, even though the human energy involved in the construction and maintenance of public facilities was provided by council employees. No farmer regarded such expenditure of his or others' time as wasted. The general view was that each must do a fair share for the benefit of all and each must take a turn on the committees so that no one person was unfairly penalised in respect of time-allocations as between public responsibilities and private farming interests.

Thus with respect to infrastructure maintenance, development has brought increased time and energy demands, with the main energy source used by farmers to supplement human labour being animal power.

6. 'Members are kindly requested to grab anything he thinks is good enough for the Show – while there are plenty farm produce, here and there'! Msengezi West ICA Committee, Minutes, 7 March 1972.

Production Energy

Cropping   An average of some 14 hectares was under cultivation on Msengezi farms in the mid-1970s, and at the height of the growing season on farms *A* and *B*, 46.3 per cent and 19.9 per cent of all labour-time was devoted to field-crops. Competition for available labour-time starts even before land preparation, in the amount of time spent on obtaining inputs. Here we may contrast farmer *A*, who had a telephone and regarded himself as 'grown-up . . . able to do my own marketing' and therefore not a member of the Msengezi Producers' Co-operative Society, with farmer B, who was a member. *A* ordered his fertiliser, pesticide and veterinary requirements direct from sales representatives visiting his farm. These inputs were then delivered to his farmyard. But for his seed he (or his adult children) ventured as far afield as Kadoma and Rusape (the latter some 250 km from his farm). In contrast, the time *B* saved on obtaining inputs via the co-operative, which he collected by scotchcart from the depot, he used to market dairy produce and fresh vegetables. Another trade-off characterised *A*'s decision to kraal his cattle at night instead of leaving them in one of his three paddocks, thus requiring additional herding labour: the crop-yield benefit from a manure-fertiliser mix on sandveld soil was regarded as outweighing the labour-time involved. *B*, who cropped only in order to provide stockfeeds for his dairy and pig enterprises, left his beef cattle in one of his eight paddocks overnight, thus cutting out the extra herding labour, and used only the manure from his dairy cows and pigs.

B's commoditised vegetables were irrigated by a diesel pump from his farm dam, whereas *A*'s vegetables, for domestic consumption, were handwatered from the manually-operated homestead well a few metres away.

The 66:34 ratio between ox- and tractor-ploughing in Msengezi, noted earlier, was assisted to a very limited extent (1.4 per cent of the total cultivated area) by the use of tractors owned by neighbouring white farmers, but ten-elevenths of all contract tractor-ploughing was done by Msengezi tractor-owners who, of course, also tractor-ploughed all of their own fields. Nearly two-fifths of local tractor-owners flatly refused to undertake any contract ploughing in an attempt to keep their machines in operation for as long as possible,[7] and nearly half of all contract tractor-ploughing was

7. They included 12 of the 20 tractor-owners ploughing less than the average 14 ha in Msengezi, who collectively ploughed an average of 9.3 ha and whose cropping strategy was intensive rather than extensive.

done by five farmers. They sometimes had to plough their own fields later than they wished, because of their non-tractor-owning neighbours' demands for tractor-ploughing. Apart from the tractor-owners, three categories of farmer availed themselves of tractor-ploughing to a level above the average: women land-owners, women farm managers and polygamists, whose average area under cultivation was nearly 150 per cent of the Msengezi average (Cheater, 1981: 368). In contrast, neither of the two case-studies chose to save human and animal draught energy by large-scale tractor-ploughing (Table 10.3). Yet clearly, investing in a tractor would not only save human energy in short supply (a motivation emphasised by Matlon et al; 1984), but would also generate cash income.[8]

However, more Msengezi farmers were planning to sink bore-holes, buy trucks or invest in diesel-powered grinding mills than were hoping to buy tractors. The scarcity of tractors in reasonable working order, the difficulties of repair, and the costs and availability of fuel and competent tractor-drivers, explain most of this paradox, to which should be added that a tractor emits not valuable organic manure but noxious effluvium (George, 1985: 93) and makes no contribution to agricultural output. And with respect to mechanical tillage, even the tractor-owners chose manual or animal-power over tractor-power for many post-ploughing operations, depending on the stage of growth of the crop; the water content of the (largely sandveld) soils; and the precision of the work required (for example, in applying carefully measured, scarce, nitrogenous top-dressing to growing crops).

If ploughing was one-third mechanised and land-cleaning mainly manual, harvesting (of all crops) used only human labour. While casuals reaping maize were normally paid in grain, the cotton harvest provided many opportunities for cash-remunerated piece-work, especially to those in the nearby communal land – but at half the rate offered by neighbouring white farmers. So there was, every year, a problem in getting all the seed-cotton in before the date by which plants had to be destroyed as part of the national pest

8. Tractor-ploughing charges varied slightly. Very close kin sometimes got away with meeting only the costs of fuel and oil, but in most cases paid the going rate of R$1.6–1.8 per ha, a charge which many Msengezi tractor-owners complained did not in fact cover the real costs of deploying their machinery but which seems to have been set by neighbouring white farmers for ploughing in areas like Msengezi. My figures suggest a total income of over R$7,500 earned by contract tractor-ploughing, 46.6 per cent of which was shared by five operators, while 37.9 per cent went to the 'big three'.

Table 10.3 Case-studies: livestock resources and production methods

Farm A	Livestock owned (early 1974)	Farm B
96	Total cattle, composed of:	32
0	bulls	2 (Jersey, Sussex)
49	cows	11
22[a]	oxen	4
25	'calves'/immature	15
7	head sold in 1973	4
7.3	1973 sale offtake as % of 1974 herd	12.5
	Other livestock:	
0	pigs	20[b]
c.40[b]	sheep	0
c.40[b]	goats	0
40+[b]	chickens	20
	Energy-related choices	
Yes	farm ring-fenced and gated	Yes
3	cattle paddocks	8
Entire herd	night-kraaling	Cows only
Council-run dip	disease-prevention	Own hand-spray

a. Twelve were split (4 + 8) between two other farmers for training.
b. For sale, not domestic consumption; excepting the pigs, women's investments.

control system. Msengezi soils are unsuitable for cotton, and yields are low: farmer *A*, for example, expected 2.7 tonnes from his 4 hectares of cotton. Msengezi farmers could not afford to pay harvest labour more, and so lost out in the competition for scarce human labour. Leaving aside the availability of machines and problems of maintenance, and irrespective of their negative effect on cotton grade and price, at these yields the mechanisation of the cotton harvest – like the mechanisation of so much else in small-scale commercial farming areas – was simply economically unfeasible.

It is noteworthy, too, that the major cost of *B*'s independent production and marketing strategy was an increased demand for human labour in production, even though his total area was half the Msengezi average. Maize (planned to yield at well over 3 tonnes per hectare), groundnuts and sunflower were grown purely for domestic consumption and for stockfeeds; the marketing of any windfall surplus in an exceptionally good year was never planned. For *B*, the

215

financial secret of farming success lay, first, in processing crop energy through animals and secondly in specialist crop lines, never in field-crops. *B*'s energy input strategies, therefore, differed from *A*'s, as we have already seen in his use of the co-operative for input supply, of cattle-kraaling and of a water pump for vegetable irrigation: all strategies designed to save human time and energy.

In addition, farmers *A* and *B* applied different levels of investment in energy inputs (notably in fertilisation strategies and rates, but also in field preparation) as between crops destined for the market and those for on-farm consumption.

Farmer *A* had faced a technical problem in planting his 1973–4 cotton crop, which resulted in another example of energy-saving innovation. His young grandchildren (visiting the farm during the school holidays) had taken two and a half weeks to plant 2 hectares and had wasted seed in having to guess where it should be planted – a problem easily solved by the purchase of an inexpensive manufactured row-mark. But that would have meant a day of farmer *A*'s own time 'wasted' in the busy season shopping in Chegutu. So, unable to sleep for thinking about the problem, he cobbled together a substitute from bits and pieces lying in the yard: a disused bicycle frame and front wheel, baling-steel strips bent into the correct interval and wired onto an old wheelbarrow wheel attached to the rear axle, the whole contraption weighed down by a foot or so of railway line and stabilised from the rear axle by 5-foot wooden struts, fairy-cycle-style. *Voilà!* A 'planting motorbike' which allowed the children to finish the next 2 hectares in five days without wasting any seed. Such inventiveness was by no means uncommon in Msengezi (Cheater, 1984: 38–9), and the inventors were extremely proud of their home-designed, home-built equipment, notwithstanding its rough appearance. If it worked in saving human labour, or using other energy resources more efficiently, no one cared much what it looked like.

Neither farmer *A* nor farmer *B* had invested in post-harvest processing equipment, though one in three Msengezi farms owned a manual maize-sheller and nearly 1 in 5 had one of a range of unsatisfactory groundnut shellers, some the personal inventions of their owners. Farmer *A*'s and *B*'s basic equipment remained remarkably similar, notwithstanding their different production and marketing strategies. However, differenc s in energy technology between the two farms were more apparent in their livestock production.

The labour requirements for tending livestock as between farms *A* and *B* somewhat paradoxically were proportionately heavier for *A*'s mainly beef herd than the dairy specialisation of *B*, at 20.9 per

216

cent and 16.7 per cent respectively of total work done during the period monitored. There were two reasons for this. First, *A* had fewer paddocks and kraaled his entire herd every night, thus requiring twice-daily herding labour to move the cattle; whereas *B* kraaled only his cows, conveniently close to the dairy for morning and evening milking. Secondly, *A* dipped his herd weekly at the nearest council-run diptank, about 1 km from his homestead, whereas *B*, three times that distance from the nearest diptank, hand-sprayed his herd on his farm, thus saving, as he put it, time, public roads and his cattle's condition. He had followed this regimen for some 15 years, under special dispensation from the veterinary services department. (Three other farmers had built or were building their own diptanks to achieve the same ends.) Both farmers *A* and *B* had cows in milk and, from their returns, seemed to spend much the same time each day in milking, perhaps because *B* strip-milked his entire herd and then fed the calves manually, while *A* took only small amounts of milk from (a larger number of) lactating mothers. Farmer *B* used five man-hours per day in milking: a more sophisticated milking technology was desirable but economically non-viable. No data were available on milk yields from *A*, but off-farm sales were limited and occasional, nowhere near the documented levels of farm *B*. However, the women of farm *A* spent a couple of hours a week preparing properly soured milk, mainly for home consumption.

From his dairy and vegetable enterprises farmer *B* received a steady, if small, cash income, which centred on his contracts to supply fresh milk and vegetables to the primary and secondary schools at the nearby mission during term-time. Farmer *B* (together with other farmers in Msengezi dairying for sale on a smaller scale) contradicts the contemporary popular (and government) belief that small-scale, localised, commercial dairying did not exist before Independence. Moreover, in those early days such dairying was initiated and organised by the farmers themselves, not by state marketing institutions in conjunction with Scandinavian aid agencies offering financial and organisational assistance to implement their own nationally developed technologies.

Modelling Energy Flows and Interrelationships in the Msengezi System

The individual Msengezi farm can be conceived as an energy system lying at the centre of two other, encompassing, energy domains: that of Msengezi and of the nation respectively (see Figure 10.1). The farm energy system is divided into domestic, infrastructural, livestock and cropping sections, each potentially having links with the local and national domains. Within the energy economy of the individual farm, bi-directional flows are theoretically possible among all four sectors. In particular, the domestic sector returns energy to the other three, as required, via the provisioning of human labour (though this is not indicated on Figure 10.1). In Msengezi in the mid-1970s the transfer of energy from the domestic to other sectors varied, with the infrastructural sector absorbing least; while more substantial reciprocal flows linked the livestock and cropping than any other quadrants. Meanwhile, of the four, the boundary between public and private in Msengezi was least permeated for the domestic quadrant and most in the infrastructural sector, as indicated on Figure 10.1.

Although a gender-based division of labour was not an ideological component of the Msengezi energy system, in practice, as we have seen, women did most of the work in the domestic sector, while also providing a significant proportion of all labour inputs to cropping, especially on polygamists' farms, and tending livestock, occasionally even herding cattle down to the public dips and competing (in mixed and all-female teams) in ox-ploughing competitions. But women did not have much input into the infrastructural sector. Nor (in any case known to me) were they the inventors of new equipment or techniques designed to save farm labour, despite often being the sole or joint policy-makers regarding production strategies (Cheater, 1981; 1984: 115–16). Similarly, while men dominated the infrastructural aspects of energy provision, they also contributed to all other sectors, including the provision of domestic energy.

Let us now examine further the significance of the farm boundary for each energy quadrant. With the exception of purchased foodstuffs and lighting fuel, the domestic sector of the farm economy was generally self-provisioning in respect of its energy requirements, including staple foodstuffs – grains, vegetables, eggs and meat – as well as water, woodfuel, and the equipment and human energy required to deliver these to the kitchen door. Farm families seemed to strive after autarky in respect of their domestic energy requirements, and where such energy was provided from beyond the individual farm, it came mainly from outside Msengezi: petroleum products, manufactured cast-iron stoves, wheat and rice products, sugar, electricity. Important exceptions to this generalisation lay in every farm's use of local mills to grind grain and the

218

Figure 10.1 Energy inputs delivery system in Msengezi

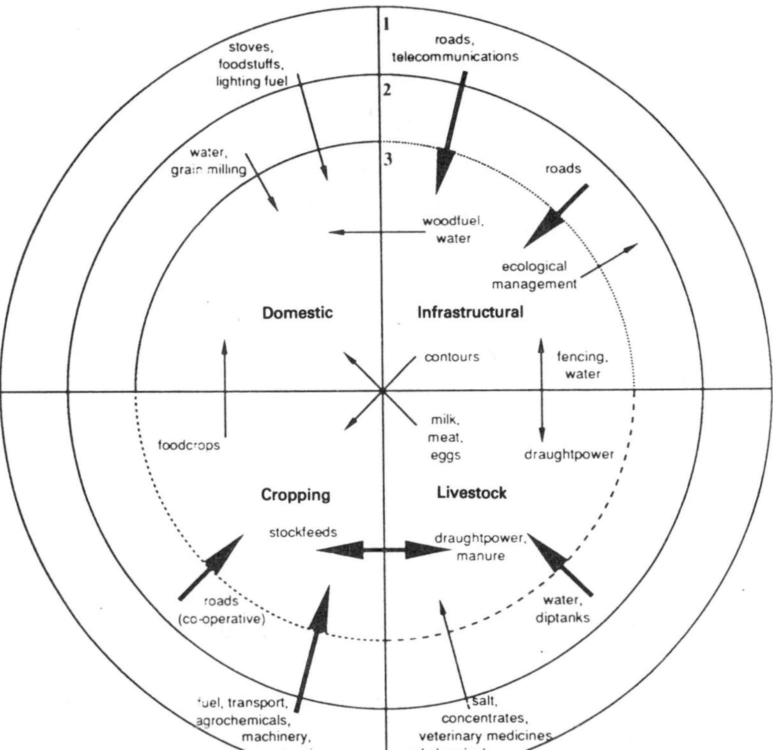

Key: 1. National domain.
  2. Msengezi domain.
  3. Individual farm.

minority reliance on public water supplies. Perhaps this pattern explained, at least in part, why women stayed on–farm for a greater proportion of their time than men: they had fewer energy interests to service in the local public domain and, therefore, did not need to devote so much time to constructing and maintaining off–farm social networks. (One should note here that the autarkic domestic energy sub–system did cover hired workers, but usually independently of their landowning employers: workers lived separately and commensality was rare, being replaced usually by rationed grain and fuelwood.)

In contrast, most of the energy devoted to infrastructural affairs in the local public domain was provided by male land-owners (and council employees). In this sector, farm boundaries merely marked a change in the locus of responsibility for facilities essential to the entire commercial farming enterprise. Outside the farm gate, everyone in Msengezi had an interest in the condition of the public roads. Inside the farm gate, only the individual farmer was concerned about the transport of his crops to market. But gullies and drain-spills were no respecters of farm boundaries! What looked like a clear division in legal responsibility was in practice a fuzzy interdependence linking individuals and local council, mediated largely by the members of the ICA committees, with their legal responsibility for both domains. It was within the framework of ICA-organised field-days that new technologies for ecological repairs and maintenance, both on- and off-farm, were disseminated.

Another major reason for the fuzzy boundary in respect of the infrastructural sector lay in the council provision, in the public domain, of part of the essential technology for raising livestock – the tanks of chemically treated water through which all cattle were legally required to take a weekly swim in the ritual called 'dipping'. There were five council diptanks, spread evenly among the seven wards to ensure that potentially damaging cattle hooves travelled the least possible distance over council roads. But while those cattle living close to a public diptank might complete dipping in an hour, those most remote would take half a day or longer to return to their farm. For many farmers, too, it was necessary to water cattle at one of the nine large dams or two boreholes constructed on public land by the council: either because their domestic water supplies did not stretch to include large numbers of livestock (the average size of farm herd in 1973–4 totalled 25 head of cattle, plus small stock among a minority, and poultry); or because their small farm dams (owned by one in every four farmers) had temporarily dried up. Livestock as a source of energy were thus equally dependent on public and private domains, on the former for their health and on the latter for their subsistence (using crop residues and hay as well as whole grain). Cattle were also dependent, for their own health and ability to deliver draught power and manure into the cropping sector, and capital into the farm budget on their final disposal, on the national provisioning system, for salt, concentrates, arsenical compounds for dipping and veterinary medicines.

Most of the externally obtained energy inputs – hybrid maize, soyabean and vegetable seed; delinted cottonseed; improved varieties of groundnuts and sorghum; chemical fertilisers; agrochemicals generally; traction power; petroleum fuels; vehicles and other equipment – were supplied from the national domain. The cropping sector, like the domestic sector and in contrast to the other two, relied on the local public domain mainly for ease of transportation rather than any direct input to the on-farm energy system. What cropping energy inputs and energy were not autarkically generated on the individual farm (i.e. manure, retained seed for minor food crops, family labour, animal power) came mainly from beyond Msengezi (including hired labour). The main exception to this generalisation was the use of tractors (for less than one-sixth of the total area ploughed) hired overwhelmingly from other farmers in Msengezi.

With respect to external inputs into the different sectors of the on-farm energy system in Msengezi, then, it is quite clear that the local and national energy domains were differentially important. While strategies for social reproduction emphasised domestic energy autarky, commercialised cropping required an ever-increasing dependence on the national domain. Infrastructural maintenance and livestock production, in turn, relied mainly on the local domain for their support. As this differentiation developed, one factor emerged as overwhelmingly significant to all domains: transport.

Conclusions

The differentiation of these sectoral energy links between individual small-scale commercial farms, the local community and the national provisioning system is important in understanding the energy resource requirements for further technological development in areas like Msengezi. The 1970s (and probably 1980s) energy priorities of most Msengezi farmers would not have included woodfuel replacements (offtake from their own supplies being eminently sustainable at 1974 levels), nor the provision of electricity. They did include a better public transport system, for people as well as marketed crops and livestock; more, cheaper fertiliser; and greater on-farm water storage capacity (for livestock and irrigation); all of which, together with new hybrids less susceptible to drought conditions, are essential for the further intensification of cropping and animal husbandry and have regularly appeared, for at least three decades, on the agenda for the annual conferences of the small-scale farmers' union.

The most fundamental problem involved in realising such priorities lies in agricultural pricing policy. Although Zimbabwe's pricing system is based on 'capitalist' rather than 'peasant' agricultural costing, it assumes either that economies of scale will be possible, or that exploitation of family labour will yield an attractive return. Small-scale commercial farmers fall between these two stools. The majority appear to be stuck in a 'middle-level equilibrium trap' which holds output on a plateau mid-way between fully commercialised and more properly peasant producers based on an 'intermediate' mix of energy sources and inputs. Their position perhaps accounts, at least in part, for their technological creativity within the energy system of the individual farm.

References

Beijer Institute, *Policy Options for Energy and Development in Zimbabwe*, Vol. 1, Main Report, Stockholm, 1985

Bourdillon, M. F. C., Cheater, A. P. and Murphree, M. W., *Studies of Fishing on Lake Kariba*, Gweru (Zimbabwe), 1985

Cheater, A. P., 'Aspects of Status and Mobility among Farmers and their Families in Msengezi African Purchase Land', *Zambezia* 3, 2, 1974, pp. 51–9

—— 'Co-operative Marketing among African Producers in Rhodesia', *Rhodesian Journal of Economics* 10, 1, 1976, pp. 45–57

—— 'Small-scale Freehold as a Model for Commercial Agriculture in Rhodesia/Zimbabwe', *Zambezia* 6, 2, 1978a, pp. 117–27

—— 'Bond Friendship among African Farmers in Rhodesia', in A. Argyle and E. Preston-Whyte (eds), *Social System and Tradition in Southern Africa*, Cape Town, Oxford University Press, 1978b

—— '"Keep your fingers out!" Some observations on financial abuse at the local level', *Zimbabwe Journal of Economics* 1, 2, 1979a, pp. 172–9

—— *The Production and Marketing of Fresh Produce Among Blacks in Zimbabwe*, Salisbury (Harare), University of Zimbabwe Press, 1979b

—— 'Women and Their Participation in Commercial Agricultural Production: the case of medium-scale freehold in Zimbabwe', *Development and Change*, 12, 3, 1981a, pp. 349–77

—— 'The Social Organisation of Religious Difference among the Vapostori weMaranke', *Social Analysis* 7, 1981b, pp. 24–49

—— '"When is a Class not a Class?" Class formation in the purchase lands of Zimbabwe', in J. Bardille (ed.), *Class Formation and Class Struggle*, Morija (Lesotho), n.d. [1981c]

—— 'Formal and Informal Rights to Land in Zimbabwe's Black Freehold

Areas: a case study from Msengezi', *Africa* 52, 3, 1982, pp. 77–91

—— 'Cattle and Class? Rights to Grazing Land, Family Organisation and Class Formation in Msengezi', *Africa* 53, 4, 1983, pp. 59–74

—— *Idioms of Accumulation*, Gweru (Zimbabwe), Mambo Press, 1984

—— 'Anthropologists and Policy in Zimbabwe: design at the centre and reactions on the periphery', in R. Grillo and A. Rew (eds), *Social Anthropology and Development Policy* (ASA 23), London, Tavistock, 1985

—— 'Fighting Over Poverty: the articulation of dominant and subordinate legal systems governing the inheritance of immovable property among blacks in Zimbabwe', *Africa* 57, 2, 1987, pp. 173–95

George, S., *Operation Flood: An appraisal of current Indian dairy policy*, Delhi, Oxford University Press, 1985

—— *Dairy Development in the Small-scale and Communal Farming Areas of Zimbabwe: a field report from three areas*, Contribution to a consolidated report on the Dairy Development Programme presented to Norad. Ms., 1988

—— and Kuimba, S., *Development and Participation: the dairy development programme in one village of Chikwaka Communal area* Report submitted to Norad. Ms., 1989

GTZ, *Energy Programme: Zimbabwe* (Expert Report), Heidelberg, 1987

Hancock, D., Katerere, Y. and Moyo, S., *Rural Electrification in Zimbabwe*, Harare, ZERO/Development Technology Centre London, 1988

Hosier, R. H. et al., *Zimbabwe: Energy planning for national development*, Stockholm, 1986

—— (ed.) *Zimbabwe: Industrial and commercial energy use*, Stockholm, 1988a

—— (ed.), *Energy for Rural Development in Zimbabwe*, Stockholm, 1988b

Matlon, P. et al., *Coming Full Circle*, Ottawa, IDEC, 1984

Temane, B., Moyo, S. and Katerere, Y., *Integrated Rural Development and Energy: Planning and implementation – issues of concern for rural electrification*, Zero Technical Paper No. 1, Harare, 1987

Zimbabwe Energy Resource Organisation (ZERO) 1988 Report on the Seminar on Rural Electrification in Zimbabwe, 26–28 July 1988, Harare

11

Choice of Technology in Food Processing for Rural Development

Domien Bruinsma and Robert Nout

Developments in Food Processing

People have always endeavoured to process food products; to make them edible, to preserve them and avoid losses, to extract certain valuable components, to produce composite products of improved nutritional or organoleptic quality, or to generate income and employment. In addition to the existing, traditional methods of food processing, industrial processing has been introduced in sub-Saharan Africa during the last decades. Often this has had important effects on rural areas. Due to the introduction of large-scale production on estates and processing of palm oil and sugar, many people were displaced from their land. At the same time, these plantations and industries created new employment opportunities, though not always in sufficient quantities to employ all the landless. This chapter examines both the potential and the pitfalls of small-scale, rural, food-processing industries. Cash-crops are not included for two reasons: first, the processing of these crops is determined more by export than by local requirements, and options are therefore often quite restricted; and second, the processes themselves are well understood.

Medium- and large-scale agro-industries had various indirect influences on rural development, especially those that were created to process local raw materials, which did not need to be grown on estates. These factories created a demand for certain raw materials and encouraged farmers to produce more cash-crops. Sometimes this reduced their available hectarage and time for food production. Although most of these agro-industries produced for the export market in the first instance, their production was often also intended for the local market and as such competed with traditionally processed food products. The medium- and large-scale food industry in Africa is now to a large extent producing for the local market. Presently, this industry is encountering various problems stemming from its 'technical' characteristics, including excess capacity, lack of spare parts and lack of adequately trained personnel, in addition to its social impact. The firms tend to disrupt the existing rural social structure by displacing female labour, introducing market relations between producer, manufacturer and consumer, and changing the ownership distribution of the means of production.

These problems have stimulated the interest of various organisations in the development of 'appropriate technology' in two directions. The first can be grouped under the heading of improving or upgrading existing processes or products in medium-scale food-processing industries. The second direction is related to household production (often by women) with the same purposes. Examples of both are given below. Special attention is paid to the social implications of those technologies that imply a shift (of at least some parts of the processing cycle) from household production to the more formal production at a larger scale.

Although there is no universally accepted definition of 'appropriate technology', and different authors emphasise different aspects of appropriateness, there appears to be broad agreement as to the main focal points, notably, the use of labour, the extent of investment, the degree of sophistication, the scale of production and awareness of the needs of the target group. The range of traditional food-processing industries at the village level is vast. But if so much is going on already, what can development agencies do and, perhaps more important, why should they get involved if it is already done quite well? A second and equally important question is: What can be done? The answer is, technically, just about anything. But there are many other factors (e.g. political, social, cultural) which can severely inhibit freedom of action. It must be realised that a piece of machinery can only be appropriate if both time and place are right. If the setting changes, or the equipment is operated in a different environment, its characteristics may no longer be appropriate.

A number of small-scale food-processing technologies have been developed during the last twenty years. The following examples are illustrative; they describe the technology, and its nutritional, technical and socio-economic advantages, as compared to traditional processing. Where possible, some assessment of the impact of the technology (both intended and unintended) and the causes of that impact are also given.

Cereal Processing

In the field of cereal processing, flour milling technology and manufacture of weaning food formulas will be discussed with particular reference to the scale of operations and the choice of processes, respectively.

Flour Milling

The scale of operations in rural flour milling has nutritional and quality, as well as economic and social, implications. Locally-grown cereals in sub-Saharan Africa include maize, rice, sorghum (guinea corn) and various types of millets. Traditionally, these are consumed as cooked doughs, pancakes, fritters, gruels (watery porridges) and beers. In many regions, cereals are soured by natural fermentation prior to cooking. The resulting characteristic sour taste is popular, and it is a common belief that fermented cereal products are more nutritious than non-fermented cereal products.

Flour is traditionally prepared by the labour-intensive process of manual pounding with pestle and mortar, combined with winnowing to remove the fibrous parts (testa), especially of millets and sorghum. On average, 25 kg of raw material could be processed daily per mortar, serving the needs of 25–35 consumers (Table 11.1). This home-scale technique has the advantages of low investment, short distribution distance and flexibility.

In addition, it is suitable for processing the entire range of available cereals. However, the task of pounding requires considerable time and therefore competes with other productive activities. Consequently, the demand for more convenient flour milling methods has resulted in the gradual introduction of the phenomenon of the service mill (Table 11.1), preceded in some areas by hand-operated maize mills. A typical village service mill consists of a 'disc' or 'plate' mill (e.g. the 'Premier Mill'), commonly driven by a 2–5 hp diesel (e.g. 'Lister') or electric engine. Payment for such custom milling is either in cash or in kind. A recent survey in Cotonou (Benin) illustrates the widespread use of the relatively small-scale technology. In this town of 420,000 inhabitants, 650 service mills are in operation with an average individual daily throughput of 215 kg (mainly of maize). Likewise, it was estimated that in Niger, in 1982, more than 1,500 service mills were in use. The widespread adoption of this technology may be due to the fact that it meets a basic felt need at an acceptable cost; that is, that the benefits (time

Table 11.1 Small-scale flour milling technologies

Technology	Scale	Throughput (kg/day)	Consumers served (number/day)
Pounding (mortal and pestle)	Home–Quarter	≈ 25	25–35
Service mill (disc or plate mill)	Quarter–Village	200–300	250–500
IDRC–RIIC (dehuller and hammer-mill)	Village–Town	2,000–4,000	5,000–10,000

and energy saved) exceed the costs (payment for service milling, distance to the mill).

Small hammer mills, however, are not immune to the problems experienced by medium- and large-scale processors. In the Equatoria region of the Sudan mechanical diesel mills stand idle because of irregular supplies of diesel, the frequent need for repairs, lack of spare parts and management difficulties (Ballot, 1985). Women's co-operatives are proposed as a management solution, and efforts have been initiated by a women's self-help project funded by the German federal government, but are constrained by the lack of traditional co-operation mechanisms between women. Both gender and ownership issues affect, and are affected by, the introduction and choice of scale of food-processing technology. In Senegal diesel mills operate efficiently and profitably in the larger and/or more prosperous villages, but the remaining poorer and/or smaller villages with smaller quantities of raw material to process may be more appropriately served by an animal traction mill because of its smaller scale and lower capital and operating costs. The women could bring their own donkeys to provide the animal draught power. A niche frequently exists for businesses of various scales using different technologies in different parts of the same country (Hyman, 1989). However, whereas animal draught power is a traditional technology in Asia, in some parts of sub-Saharan Africa it is not, so its appropriateness in a particular situation would have to be evaluated in the same way as that of any other introduced technology.

The quality of the products processed by these village mills is crude, since the entire cereal grain is ground without separation of the fibrous seed coat. This need not be a disadvantage when maize is milled since its weight fraction of seed coat is relatively small. In fact, whole maize meal tends to be more nutritious because of the oil and vitamin content of the maize germ it contains. On the other hand, sorghum and millets have a much larger proportion of seed coat containing indigestible fibres, tannins and pigments. When ground to a meal in a single disc-mill, they yield unacceptable flours. The production of sorghum and millets represents approximately 20–25 per cent of the total African cereal crop. There is a tendency to regard them as 'poor man's crops', since they are mainly processed and consumed on a domestic scale in the rural areas. Sorghum and millet are better suited to semi-arid conditions than maize, and the growing of a diversified cereal crop which includes them may be of strategic importance to national food security, as well as contributing to development objectives that favour the rural population. The growing consumer demand for imported grains and flours (rice, wheat and maize) is unsustainable (Delgado and Miller, 1985). During the 1970s, efforts were made to promote the growing of more sorghum and millets, and to introduce high-yielding and bird-resistant new varieties. However, there was no incentive for farmers to produce more than the subsistence level, mainly because urban service mills were unable to process surplus sorghum or millet into acceptable consumer products. Furthermore, with the expansion of maize mills into marginal areas, better suited to millet and sorghum production (e.g. parts of eastern Kenya and northern Zambia), women have experienced an inducement to shift both crop production and consumption towards this less appropriate crop, for which hand-pounding is no longer necessary. Development of domestic or village-scale sorghum and millet dehullers could help to contain this trend (Cf. Haswell herein).

The creation of an urban market for surplus sorghum and millets was made possible by the introduction of a technology developed in response to the needs of rural women to reduce the time spent in food preparation and processing. A new system for small-scale dehulling or decorticating and flour milling was developed by the IDRC (International Development and Research Centre, Canada). The basic design of the dehuller (Figure 11.1) consists of a metal shaft on which a number of grinding stones are evenly spaced about 2 cm apart. This rotor is enclosed by a barrel which is partly filled with grain. The discs run at 1,500–2,000 rpm and gradually wear off the outer layers of the grain. (Legumes can be processed too.) Subsequently, the dehulled grains are ground to flour using a hammer-mill. Several prototypes were tested: the early Botswana-made 'RIIC' (Rural Industries Innovation Centre) dehuller performed only on a continuous flow basis, and could not deal with

Figure 11.1   The basic design of the dehuller

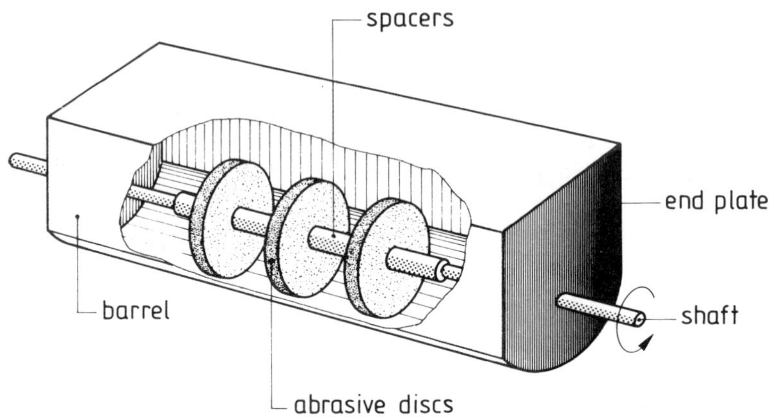

small batches being brought by individuals who were seeking a milling service for 'their' sorghum. A needs reassessment led to successful modification. The scale and capacity of the RIIC dehuller, while appropriate to Botswana where 80 per cent of the population live in large villages, was reduced to fit better the sparser settlement patterns in The Gambia and Zimbabwe (Schmidt, 1988). Thus, considerations of capacity utilisation, distance of users from the 'service', and settlement pattern influenced the appropriate scale of the dehuller and adapted versions have been locally produced and are operational in Botswana (37 in use in 1987), The Gambia (13 in use in 1987) and Zimbabwe (40 foreseen in 1990). The present RIIC dehuller can handle batches as small as 5 kg and, in continuous flow operation, can process 5 tonnes per 8-hour shift. The economical breakeven point lies around 1.5–2 tonnes per day. From Table 11.1 it can be seen that this scale of operation can only be profitable under semi–urban conditions, considering the throughput, cost of distribution and the fact that it produces a product (sorghum and millet flour) which has to compete with the usual staple (maize or rootcrops).

The IDRC experience was summarised by Schmidt (1987) as follows:

1   Economies of scale apply to the marketing and distribution of the processed product, but not to the processing itself.
2   Investments were not as profitable as expected because of heavy competition between mills and oversupply of installations.

3   Credit agencies had not anticipated occasional droughts and consequent reduced throughput.

Nevertheless, machine dehulling would merit a prominent place in African cereal processing, since:

1   It frees woman and child labour for other tasks.
2   It successfully handles high tannin (bird–resistant) cultivars of sorghums and millets.
3   It helps to generate urban demand for sustained surplus production.

In Botswana despite efforts to direct the technical capability of the dehuller towards rural beneficiaries, and to steer the dissemination of the technology towards group investors such as co-operatives and development trusts, rather than individuals, little progress was made towards the achievement of either of the development objectives – rural job creation or a decentralised milling industry. The nature of the industry's evolution, which was influenced by an ambiguous policy environment and critical raw materials shortage caused by drought, may have caused a permanent change in the way sorghum is processed; commercial milling has almost completely displaced service milling (Schmidt, 1988). No attempt has been made to evaluate the impact of the ownership distribution of the dehuller technology, though other studies indicate that ownership and control of food processing technology may be a key factor determining its sustainability and differential impact (ILO, Geneva, 1984; Hyman, 1989).

Weaning Food Manufacture

Weaning foods are used to feed infants of 6–24 months, especially when breastfeeding needs to be supplemented, but children are still too young to eat the family diet. In urban and rural situations, a growing demand for convenient weaning food has developed for similar reasons as outlined for cereal flours. In addition, the specific nutritional needs of small children are often not met by the traditional porridges based on staples such as maize or cassava. Protein-Energy-Malnutrition (PEM) and watery diarrhoea contribute significantly to the prevailing high infant mortality rate. Adequate weaning foods must therefore supply sufficient energy (4.2 kJoule per gram of porridge) and protein (8 per cent net protein utilisation), and must be microbiologically safe.

Using locally available and acceptable ingredients, it is possible to meet these requirements by mixing staples (cereals or rootcrops) with high-protein ingredients (beans or other legumes). Sufficient 'energy-density' can be obtained with oil (expensive), sugar or malt (germinated cereal). In most African countries, market prices for cereals and legumes fluctuate widely according to season and availability. If we define the prices immediately after harvest as 'lci' (lowest cost of ingredients), it is not unusual that home-prepared weaning food costs 2–3 × lci. This is in sharp contrast to the price of weaning formulae imported from Europe or manufactured by local large-scale industries under foreign licence. Such products cost 10–25 × lci, and can only be afforded by consumers from the middle and higher-income brackets.

The Royal Tropical Institute of Amsterdam (KIT) is assisting in the establishment of rural, small-scale weaning food factories in African countries, including Burundi, Ghana, Benin and Malawi. The main features of these operations are (KIT, 1987):

1   The product can only become an acceptable alternative to home preparation if the selling price is not higher than 2–3 × lci.
2   The scale of operations should be sufficiently large to support the cost of hygienic processing facilities and labour wages; on the other hand, it is limited by the market demand, costs of transport of raw materials and final product, and capacity of storage facilities. In semi-urban surroundings, daily production of 100–200 kg finished product by 3–5 persons is regarded as feasible.
3   Ingredient long-term storage capacity is part of the operation. It enables the purchase of ingredients (at the lowest price = 1 × lci) for 60 per cent or more of a year's production capacity.
4   A simple dry process (Table 11.2), involving cleaning, roasting, dehusking, mixing, grinding and packaging, results in a sales price of 2–3 × lci.

The packaged dry mix should be cooked into a semi-liquid porridge at home. Due to constraints of time, cooking is often carried out only once or twice a day. Since the infants require more frequent feeding, e.g. 3–5 times daily, porridge is left over for the next feeding time. Under the prevailing conditions of poor hygiene and lack of refrigerated storage, explosive growth of contaminating micro-organisms takes place. Such microbial activity may easily cause gastroenteritis with acute diarrhoea.

The above example illustrates the health risk associated with

232

conventional weaning food prepared and handled under primitive conditions. At the Agricultural University of Wageningen, a method to improve the microbiological safety of cooked porridges was developed. It is based on a simple fermentation system, which can be carried out with only a few vessels, without requiring sterile conditions or pure starter cultures (Nout et al., 1989b). Fermentation is a traditionally known and accepted processing technique in Africa. Souring by lactic acid bacteria lowers the pH of the final product, and has been shown to protect it from proliferation of contaminating intestinal bacteria (Nout et al., 1989a). In addition, digestibility of nitrogen compounds and bio-availability of minerals of the porridge are improved (Lorri and Svanberg, 1988; Svanberg and Lorri, 1988). In addition, oligosaccharides from pulses which are responsible for flatulence and 'loose stools' are reduced through fermentation.

For the purpose of fermentation, a wet process (Table 11.2) is required, involving cleaning, dehusking, mixing, coarse grinding, moistening, fermentation, dehydration, fine grinding and packaging. This process at 200 kg/day is 20–40 per cent more expensive than the dry process. It also takes more processing time and involves more process operations. Present investigations aim at quantifying the nutritional benefits of the wet process; if these compensate for the higher cost involved, the total of hygienic and nutritional advantages render it a promising technique for future field testing. The successful introduction and adoption of these techniques and subsequent consumption of the product would be expected to have a favourable impact on nutrition and health. Whether rural women will be willing and able to purchase these locally-produced weaning foods will depend, *inter alia*, on their awareness of their availability and advantages, the strength of competing demands on their limited income, and on the value (in its next best alternative use) of time saved in the home-preparation of weaning foods.

Oil and Seed Processing

For the processing of oil seeds into vegetable oil, technologies for various scales of processing are available. A choice, therefore, needs to be made as to 'appropriateness' in each situation, given the social, cultural and economic environment and the priority development objectives. Processing can be done at large-scale (exceeding 1 tonne/hour) through expelling or solvent extraction. For medium-

Table 11.2 Weaning food process options. Scale of production
approx. 200 kg/day

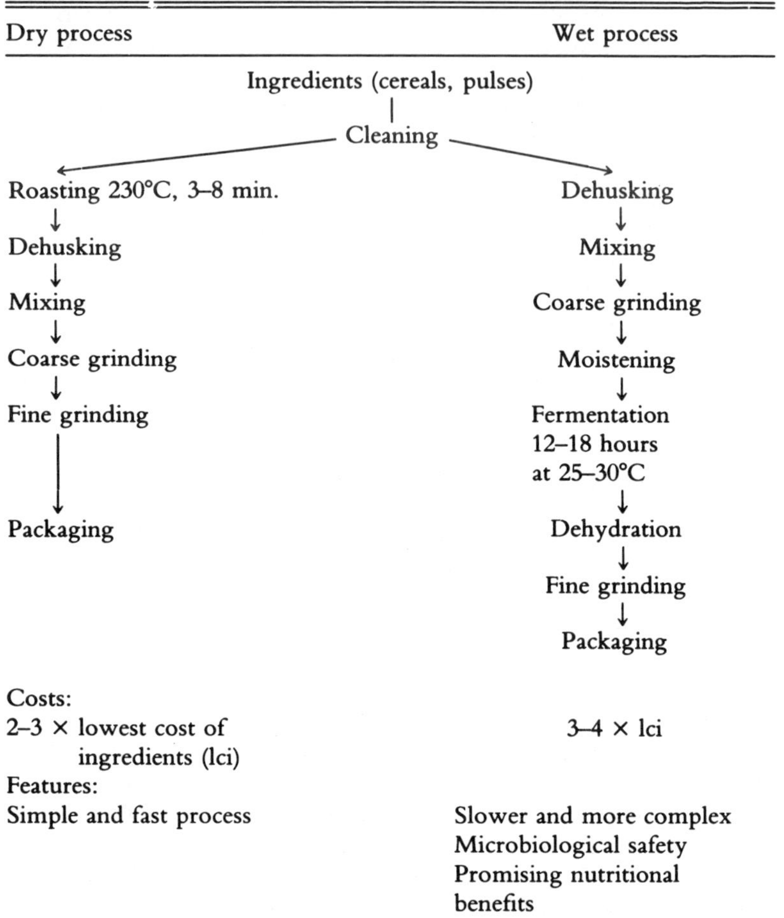

(100–1,000 kg/hour) and small-scale processing (10–100 kg/hour)
both expelling and pressing are commonly-used techniques. A problem of expellers is their wear and tear, which necessitates regular
supply of spare expeller screws or repair by welding with a very hard
alloy. Oil pressing is carried out using hydraulic presses fitted with
larger industrial jacks or even a simple lorry jack. Both these technologies have been tried out at village level. Since 1982, a project has
been set up in Mali to develop and disseminate a system for processing shea nuts that improves on the traditional system. For this project,

an oven, a hydraulic hand press and a cake-expel stand have been developed by KIT. After introductory tests, the local production of the equipment has been set up and the equipment is being disseminated.

Oil cannot be extracted from seeds in one single operation; the process consists of several stages. In Figure 11.2, the traditional (wet) process is compared with an improved small-scale 'dry' process. Size reduction to open up the cell structure, and moistening and heating are crucial to obtain a good oil recovery. The 'dry' process employs the manual hydraulic press and is an interesting option for women because:

1 The extended heating of kernels in a traditional oven can be omitted, thus saving firewood and time.
2 Work is faster and lighter because the mass is easier to pound.
3 Little water is required, which saves work.
4 Oil recovery, which of course depends on the quality of the kernels, is higher, i.e. 38.5 per cent average yield against 25 per cent in the traditional process, that is about 75 per cent against respectively 50 per cent of total fat available in the kernels.
5 The cake, which is usually discarded, can be used as fuel.

This process is likely to be technically and culturally acceptable to women, because the crucial steps, such as 'size reduction' and 'cooking', are still carried out in the traditional way, by pounding and a pot over a fire, respectively. Learning from the experiences with shea butter production, the design has been improved to prevent rapid wear of parts of the jack pump and reinforce the frame. Notwithstanding the negative experiences that called for the improvements, the demand for shea nut presses is steadily increasing. By early 1987 the processing equipment had been installed in thirty-five villages in Mali. In addition, enquiries from neighbouring countries are increasing (Wiemer and Korthals-Altes, 1989). The small-scale expeller and hydraulic press are still too large in some village situations, and require too large an investment. For these reasons, the 'scissor jack press' and the low cost 'Ram press' were developed and are being tested in Tanzania (ATI, 1989).

A system for low-cost, small-scale, oil palm bunch processing was developed in Ghana by Kramer from the 'service mill' concept (Blaak, 1989). The joint processing by farmers and mill owner minimises the potential labour displacement and loss of control effects, and maximises the profits of small-scale producers, many of them women. The owner provides a fast-running digester,

Figure 11.2 Traditional (wet) oilseed processing

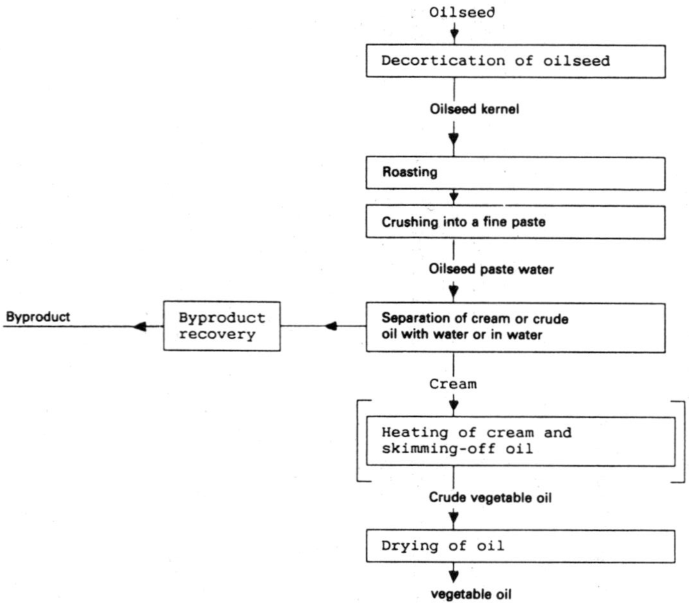

Figure 11.2 Improved small-scale (dry) processing

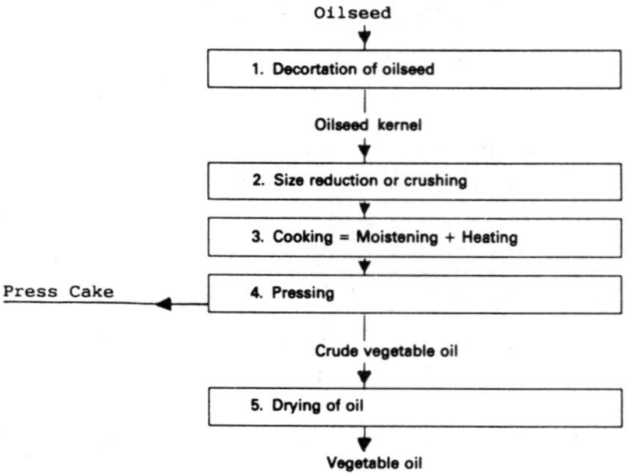

manually-operated hydraulic presses and a few operators. The farmers are responsible for bunch stripping, fruit cooking, transport of hot fruit to the digester, collection of the crude oil from the press and clarification of the oil, nut removal from the press cake, fermentation of the presscake and second pressing of the presscake. Milling charges are kept low (as the need for stocks and substantial working capital is eliminated), the identity of the fruit is maintained throughout the process, and a farmer planting higher-yielding tenera palms and harvesting at the correct ripeness receives the full benefit of her/his work. The Kramer mill employs eight persons per mill and has an oil to bunch extraction rate of only 2–3 per cent below the extraction rate of a high technology mill. Despite the presence of two high-technology mills, there are now twelve Kramer-type mills operating the Kade region, using locally-made equipment. The method allows considerable rural employment and provides a normal labourer's income to the numerous women who do most of the processing. In some areas, however, the crop density is too low to justify the establishment of even a Kramer mill, so ITA (International Technical Assistance) in the Netherlands are investigating the feasibility of a mobile mill.

Root Crop Processing

Cassava is the most important of the root and tuber crops in sub-Saharan Africa and it is widely processed both for home consumption and for sale. Cassava is the highest carbohydrate source per area, except for sugarcane, which makes it a particularly important crop given the decline in area of arable land available per person in rural areas. Per capita consumption of cassava in sub-Saharan Africa averages 118 kg/year, with a wide variation from 20 kg/year in low-consumption countries to over 400 kg/year in high-consumption countries (Hahn, 1988).

In West Africa, the market demand for cereals is growing more rapidly than the demand for cassava and other root and tubers. Probably this stems from the limited supply of cassava in urban areas, caused by marketing difficulties because of the poor keeping qualities of fresh cassava roots, and the availability of relative cheap, imported cereals (Delgado and Miller, 1985). FAO data suggest an increase in per capita consumption in parts of East Africa and highland Africa (Hahn, 1988). Bitter varieties enjoy increasing popularity with farmers since they are more resistant against field and storage pest and conditions of drought. The processing of these bitter varieties of cassava is critical. Whereas sweet cassava can be consumed fresh or be simply cooked, the bitter varieties need to be processed to remove naturally occurring toxic components. Toxicity in cassava is caused by the poison cyanide. Toxic effects occur when cyanide is liberated from a more complex chemical compound linamarin. The human body is able to detoxify the poison to a certain extent, but this goes at the expense of sulphur-containing proteins. If the diet contains sufficient protein, the body can withstand moderate quantities of cyanide. However, because of the very low protein content of cassava, accumulation of cyanide may occur with poor people having inadequate supply of protein-rich supplementary food. Both endemic goiter and cretinism can be aggravated by continuous dietary cyanide exposure from insufficiently processed cassava. The disorder *konzo*, a spastic paralysis of the legs, has also been associated with eating inadequately processed cassava and lack of protein-rich food (Rosling, 1987). These food-borne disorders have been recently reported in Zaire, Uganda, Tanzania and Mozambique. In Uganda and Mozambique it appears that the bitter varieties have been adopted by farmers before they had learned how to process the tubers properly so as to reduce the cyanide content. After having experienced the toxic effects of bitter cassava, the farming population starts experimenting with alternative processing methods, thereby often adopting the methods that are used in the area where the bitter variety came from. Likewise, cassava may be insufficiently processed during famine situations, because of lack of time to carry out the sometimes long detoxification methods. International research efforts have only recently focused on the effects of processing on the reduction of cyanide in cassava. Studies are now in progress to assess the fate of cyanide during the changeover from traditional methods of processing, their efficiency in detoxification, and the biochemical processes involved. The aim is to identify time-saving detoxification methods (Essers, 1989) for rural introduction.

Many products can be made from cassava, including fermented products such as *gari*, cassava flour for bread-making, cassava starch, as well as chips and pellets for animal feed. Each product can be made using different scales of production from household to industrial level and with a variety of equipment (Bruinsma et al., 1983), and the choices and issues involved are illustrated here with reference to *gari* production. *Gari* is a popular convenience staple food in West Africa, but similar products are known elsewhere, e.g. *farinha de mandioca* in Brazil, and *farinha torrada* or *rala* in parts of Mozambique. The *gari*-making process consist basically of

238

washing the tubers, peeling and grating, followed by fermentation during a few days. This fermentation takes place under pressure to squeeze out part of the moisture. The final step is a combined drying and frying operation aimed at preserving the product and pregelatinisation or precooking of the coarse flour.

Bruinsma et al. (1983) compared three scales of *gari*-processing operations (small-, medium- and large-scale), producing *gari* at approximately 2.2 kg/woman-hour, 70 kg/(wo)man hour and 450 kg/hour respectively. The fully mechanised process is inappropriate, given the needs, resources and management capability of a rural community. It needs a high level of investment, a well-organised supply of roots requiring one or more plantations, it employs relatively few people, it needs skilled labour and it has high costs of services (water, energy). Improvements in the traditional process (see Figure 11.3) would be much more likely to result in maintained (female) employment levels, increased labour productivity and increased output of *gari*. Generally, the first operation to be mechanised is grating. One hour mechanical grating saves 21 women working hours and is not as harmful to the hands as manual grating (Hahn, 1988). Mechanical graters are very useful but too expensive to be used at the village (less than 1,000 inhabitants), or single household level. Simple diesel-powered graters with a capacity of approximately 200 kg roots per hour can be mounted on a lorry, however. In this way, they can serve the countryside village 'factories' profitably, but have the disadvantage that they are owned and operated by men; this may imply negative consequences for the female labour force. It has been noted that where a more capital-intensive technology is introduced, men take advantage of the business opportunities it presents, and women tend to lose out (Burne, 1988). However, this is not an automatic consequence. A study of fifteen small-scale *gari*-making firms in the Brong Agafo region of Ghana illustrates the positive impact of small-scale business on the generation of employment, whether part-time, seasonal or full-time. Women entrepreneurs increased their output and incomes using inexpensive low-technology equipment; the only mechanised process was cassava grating, which was done at a service mill in the town owned and operated by men (White, 1985).

Women entrepreneurs have a potentially stronger position when they are organised in their own groups or cooperatives. This gives them better access to credit and extension advice and a better position from which to negotiate marketing contracts and to introduce new products. These co-operatives can provide a basis from which women gain control and ownership of mechanised food-

Figure 11.3   Improvements in the traditional processing of cassava roots

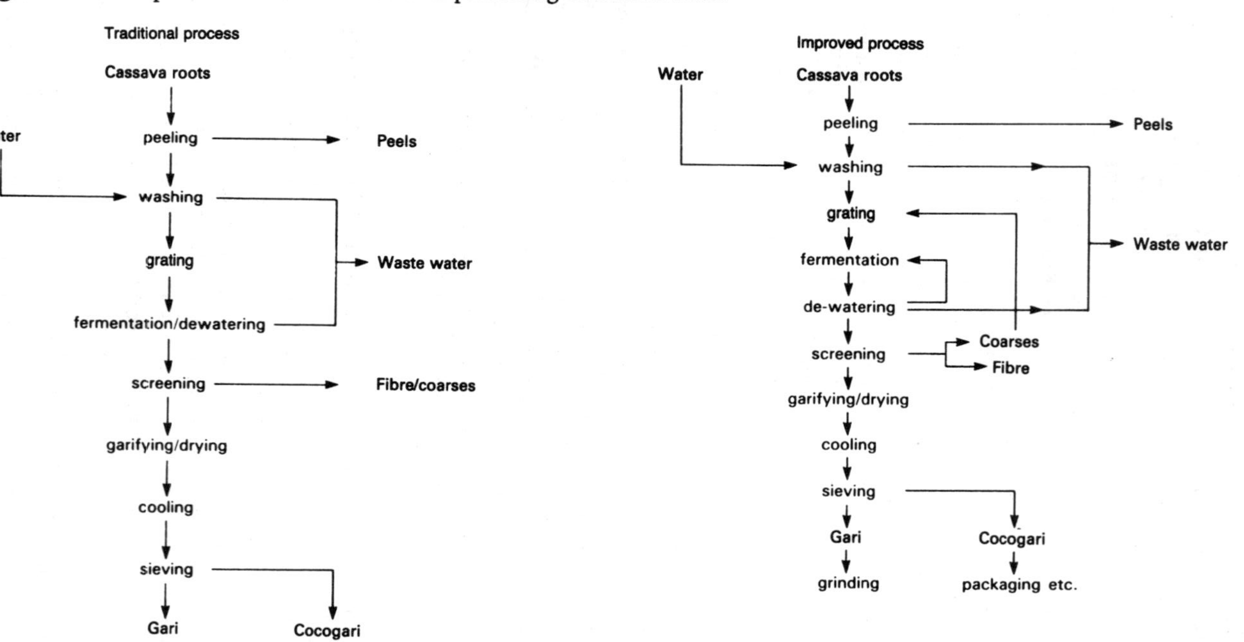

Note: The improved process can be applied with either manual or mechanised grating. The changes consist in (i) an improved fermentation process and (ii) the recycling of coarses. The former speeds up fermentation; the latter speeds up dewatering and gives a higher yield. Packaging contributes to product hygiene and ease of distribution.

processing technology. This is occurring in Ghana where a women's co-operative, heavily supported through an FAO People Participation Project, began producing and selling *gari*. Using a modified traditional technique, a communal 'roasting shed', and through joint bulk purchasing of raw cassava, the fifteen women accumulated savings and were able to obtain the necessary credit to purchase a locally-made, gasoline-powered, cassava grating machine. The project provided intensive training in accounting and business management, and the volume of *gari* produced by members continues to expand. There are signs, however, that the management capacity of the group is nearing its limit. The long-term success of the enterprise will depend on how well its members solve future business crises. This co-operative *gari*-processing enterprise has already made a positive impact. Other villagers have been motivated to undertake *gari* processing, women members of the co-operative are much more confident and there are visible signs that their living standards have improved (housing improvements, more children at school, eating a better, more varied diet). There are now two fully-employed *gari* grater operators in the village, the women have plans to use the cassava peel to feed pigs, and people making sieves for the *gari* makers have a booming business (Nyarko-Fofie, 1989).

Initiatives are also being implemented at national level. The UNICEF/IITA Collaborative Program For Household Food Security and Nutrition took the initiative to set up Rural Cassava Processing and Utilisation Centres in Nigeria to promote the production of new and better quality cassava products, demonstrate new ways of cassava utilisation and seek a successful mode of management for the centres (Kwatia, 1986).

Stimulation of Food Processing for Rural Development

Strategies to stimulate the adoption of improved food processing technologies require, first, that their goal is clearly articulated. The priority needs, the desired impact and the intended beneficiaries must all be explicitly identified, as these determine the choice and scale of technology, which in turn will have to be supported by government policies, institutions and infrastructure.

Socio-economic Factors Influencing Transferability and Choice of Scale

Food-processing technologies, which have proved to be successful in the situations for which they were developed, may not be easily transferable. Experience shows that successful introduction and adoption depend on a variety of socio-economic factors, as well as on technical appropriateness and financial viability. Technological innovation is a form of social change; acceptance depends not only on efficiency of hardware and economic soundness, but also on the degree of socio-cultural resistance (ILO, 1984). The following list of socio-economic factors is not meant to be complete, but only illustrative of some of these issues:

1 Traditional work and co-operative systems.
2 Entrepreneurship and ownership.
3 Demand for products and product quality.
4 Economies of scale.
5 Objectives.

Knowledge of traditional work habits and co-operation systems is essential in the selection and adaptation of technology aimed at saving women's labour-time. The most important processes in this respect are those of cereal dehulling and milling, and production of vegetable oil. In projects on cereal milling, women's needs and working habits are often not sufficiently known or taken into consideration in the development and introduction of new technologies. It is the experience of various projects that the closer a new technology is to current practice, and the greater the ease with which the 'new' methods can be mastered by the available skill, the greater is the likelihood that sustained use will be made of the 'new' technology. A multidisciplinary framework for analysing the effects and impact on women of technological change in food processing (ILO, 1984), and based on field work in Ghana, Sierra Leone and Kenya, assessed the importance of a range of both technical factors (e.g. auxiliary input requirements such as water and fuel, potential use of by-products), and socio-cultural factors (e.g. similarity to traditional tasks) in technology transfer. The study concluded that successful adoption is more likely when women have control over the increased income accruing, when the introduced technology does not disrupt concurrent domestic work and child-care arrangements, and when it is close to the source of raw materials.

When women's food-processing activities are mechanised and commercialised, they are frequently taken over by men. In Nigeria, mechanisation (especially at the industrial level) has been associated with the displacement of women from traditional productive activities (Ladipo and Ade Alao, 1984), in favour of men who own and operate the grating and dewatering machines used in *gari* processing (Williams, 1981). Female labour displacement resulting from the introduction of food-processing technology is reduced when the users of the technology maintain control (and ownership) over their product. Both custom hire arrangements and service mills have proved positive in this respect, as have the joint processing arrangements in Ghana (Kramer mills) which maintain the farmers' interest in the oil and enable her/him to realise a larger share of the value-added.

Greater benefits are expected to accrue to those who own and control the technology itself, and initiatives to promote and support women's co-operatives are being directed towards this end. However, it is often difficult to establish an adequate organisational structure enabling women to manage mechanical mills and to ensure the maintenance, repair and supply of spare parts (Ballot, 1985). In the introduction of labour-saving technology in the processing of oil seeds, related problems are encountered with regard to selection of an appropriate organisation structure for women's groups to manage a hydraulic press. Often the available small-scale equipment still has a too high capacity to be used profitably by groups of women, who co-operate by experience easily. The group of women, necessary to make a hydraulic press a profitable operation, is larger and thus there is a need to introduce new collaboration structures. Technological development work on the equipment for oil production is however still going on (Wiemer and Korthals-Altes, 1989). Successful business management of mechanical equipment requires a variety of entrepreneurial skills, and co-operative ownership of oil processing mills in West Africa has not worked well because of the scope for corruption (Blaak, 1989). Burne (1988) concludes that if the beneficiaries cannot own and control the improvements you are introducing, individually or at least collectively, then the benefits have to be pretty good to outweigh their losses of economic freedom and self-reliance. This brings us to the factor of entrepreneurship, which is very important in the management of processing activities. It demands specific skills to keep the equipment operating at an economic level, organise maintenance and repair, and the supply of spare parts.

The adoption of food-processing technology and the resulting impact are influenced (and possibly constrained) by the level of demand for the processed product, and also for its by-products. Demand depends on the product's price, the price and availability of substitutes, consumer tastes and preferences, factors like ease of preparation, similarity to traditionally processed products, and of course, disposable income. In some parts of Africa, the preference is for more full-flavoured, darker coloured oil made from wild palms or groundnuts in the traditional ways, rather than for the refined cooking oil (Blaak, 1989; Machell, 1989). In contrast, advertising campaigns and ease of preparation considerations have increased the demand for refined flours made from imported grain (ILO, 1984; Delgado and Miller, 1985). Although market research has not usually been part of the food technology development process in Africa, it clearly has a major role to play in defining appropriate food processing activities.

The fact that economies of scale play an important role in the profitability of a technology is well known. However, for certain processes intermediate levels of technologies are either scarce or not available. So there is little scope for profiting from economies of scale. For instance, in oilseed processing, traditional family-level processing is profitable because the opportunity cost of labour in the off-season is very low and equipment used is simple. The use of medium-scale hydraulic presses mentioned earlier, requires much organisation and co-operative processing in order to be cost-effective. The next step is often the large-scale processing factory operating at a higher efficiency. Medium-scale processing facilities operating as service mills are still scarce in sub-Saharan Africa. In addition to the competition of large-scale oil mills, medium-scale units must compete with cheap oil bought on the market through food aid programmes. Blaak's study of labour-intensive, small-scale oil palm pressing in Ghana refutes the assumption frequently made that the larger the scale, the lower the per unit production cost (Blaak, 1989). Larger-scale equipment is very costly; the main costs being interest on and depreciation of capital. In sub-Saharan Africa, labour costs are relatively low, which gives an economic advantage to labour-intensive systems of production and organisation.

Unfortunately, development agencies often work in ignorance of what their colleague organisations are up to. Sometimes this causes economic mishaps such as disturbance of local market prices; on other occasions, technological standardisation and exchange of know-how have been frustrated by uncoordinated dumpings of a confusing variety of brands, types and sizes of machinery with different origins and instructions. Frequently, the aims of separate programmes diverge and their objectives conflict. Even within the same project, conflicting objectives are formulated. Sometimes a project for food processing aims at the same time to supply income for rural women; process existing surpluses such as seasonal surpluses of fruit and vegetables; improve nutrition; respond to a named demand by consumers; and substitute imports. Of course, these aims cannot always be met all at the same time. Likewise, in projects introducing cereal mills the alleviation of women's work and the generation of income are conflicting. The costs of grinding cereals should be as low as possible to serve many women, thereby satisfying the primary objective. However, with a low price per kg it is impossible to create large wage benefits. This is the case where the target beneficiaries have not been clearly identified. Are they the rural women, the mill owners and employees, or the consumers of the processed product? It has already been stressed that the choice of a 'new' food processing technology, both of scale and of hardware, needs to be consonant with the broader development objectives of governments.

In an attempt to approach the problem of selection of technology in a systematic way, the decision scheme presented in Figure 11.4 was developed. The factors mentioned above are to be included in the criteria that determine the selection. As the figure shows, the entire selection procedure should be followed in close consultation with the target group. A final example illustrates the dangers of not doing so.

Traditionally, the production of palm oil in oil palm- and cocoa-growing areas in Nigeria is a quite labour-intensive process and results in a low rate of extraction. The men harvest the fruit, and the women crack the kernels and process the oil. The oil yield of this traditional process is only 40–50 per cent of the oil available in the fruit. Pressing the hot pulp in two stages with a hydraulic press may increase the yield to 90 per cent. The potential increase in oil yield and the necessity of diversifying agricultural production in an area prompted the Ministry of Agriculture in Nigeria to introduce the hydraulic press to a village called Gbebun. However, some women never used the press, and others stopped using it after some time, mainly because the by-products contained less oil and therefore their fuel quality was lower, and the increased oil production belonged to the men (Obibuaku, 1967). One could learn from this case-study that, although meetings were held and ideas exchanged, the villagers did not get enough time to think about what the proposed technology would mean to them. The role of the women in the traditional process was not fully recognised, and they were not sufficiently involved in choosing the technology and introducing it.

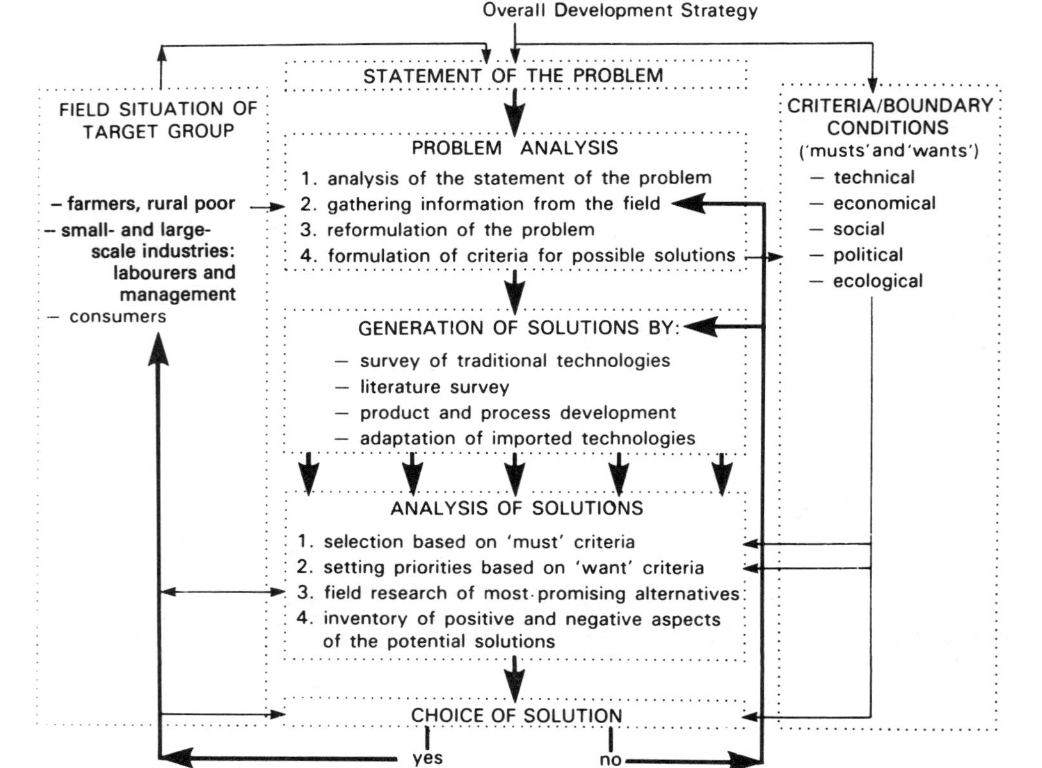

Figure 11.14 Decision-making scheme for choice of food processing technology

Potential Impact of Food-Processing Technology

The impact of food-processing technology on women may be positive when they are released from the time-consuming and arduous work of processing food by hand, and are thus able to spend more time on food or cash-crop production, on other income-earning activities, caring for children or resting. However, when food processing is mechanised, women may have to spend time and effort head-loading their grain to the mill, wage labourers may lose their means of livelihood, the value-added in food processing may accrue disproportionately to the owners and operators of the technology who are usually men, and shortages in raw materials may be exacerbated.

The longer-term impact depends on sustainability: whether the technology continues effectively and efficiently or falls into disuse when diesel or spare parts become unavailable; whether the ownership/management arrangements are sustainable: whether existing government policies (price ratios, import prices, etc.) continue and the profitability of the processing operation is maintained. The viability of women's co-operative groups as technology and business-managers, and of small-scale rural enterprise generally, depends on continued support to local technology units and centres for research and development work and help in building an indigenous technological capacity.

But alongside the recommendations, the question still remains: Is there sufficient evidence that small-scale food-processing technology is more likely to have the intended impacts than medium- or large-scale activities?

Acknowledgement

The authors wish to thank Louise Bevan, Sander Essers, Theo van de Sande and Willem Würdemann for their interest and helpful comments.

References

ATI, 'Sunflower Seed Oil Processing in Tanzania', *Appropriate Technology International Bulletin* no. 17, 1989

Ballot, J., *Food Systems and Society: The situation of women in the equatoria region, Sudan*, Report on a Mission, Working Paper 1. UN Research Institute For Social Development, 1985

Blaak, G., 'New Concepts in Low-cost Small-scale Oil Palm Bunch Processing in West Africa', Paper submitted at the International Conference on Palms and Palm Products, Nigerian Institute for Oil Palm Research, November 1989

Bruinsma, D. H., Witsenburg, W. W. and Würdemann, W., *Selection of Technology for Food Processing in Developing Countries*, Wageningen, Pudoc, 1983

Burne, S., 'Food Processing Calls for the First Steps in Technology', *Ceres*, FAO, 21, 4, 1988

Delgado, C. L. and Miller, C. P. J., 'Changing Food Patterns in West Africa. Implications for policy research', *Food Policy*, 10, 1, 1985

Essers, A. J. A., 'Detoxification of Cassava at Household Level in Rural Africa', Project proposal, Wageningen, Agricultural University, Dept of Food Science, 1989

Hahn, N. D., 'Cassava Production/Utilization and Social Change, The role of cassava in bringing positive change to African households', International Society for Tropical Root Crops Symposium, 30 October–5 November Bangkok, 1988

Hyman, E. L., 'The Importance of Small and Micro-enterprises in Rural Areas of Africa', Paper for the Expert Consultation on Small Rural Enterprises in Africa, FAO, Rome, 11–15 December 1989

ILO, 'Technological Change, Basic Needs and the Condition of Women', Report of the Joint ILO Government of Norway Africa Regional Project, ILO/NOR/78/RAF/27, Geneva, 1984.

——, *Small-scale Maize Milling*, Technical memorandum no. 7, Technology Series, second impression, Geneva, 1987

KIT, *Weaning Food, A new approach to small-scale weaning food production from indigenous raw materials in tropical countries*, Amsterdam, Royal Tropical Institute, 1988

Kwatia, J. T., *Report on the Existing Cassava Storage and Processing Technologies in Southern Nigeria with a View of Making Recommendations for the Establishment of Rural Cassava Processing and Utilization Centers*. Ibadan, Nigeria Spectrum Books, 1986

Ladipo, P. A. and Ade Alao, J., 'Mechanization of Food Processing, A Study of Women's Needs', in *Women's Contributions to Food Production and Rural Development in Africa*, Proceedings of a Workshop, AAASA, Addis Ababa, 1984

Lorri, W. S. M. and Svanberg, U., 'Improved Protein Digestibility in

Cereal-based Weaning Foods by Lactic Acid Fermentation', Paper presented at the Third Africa Food and Nutrition Congress, Harare, Zimbabwe, 5–8 September 1988

Machell, K., 'The Introduction of Edible Oil Processing in a Southern African Country', Paper for the Expert Consultation on Small Rural Enterprises in Africa, FAO, Rome, 11–15 December 1989

Nout, M. J. R., Rombouts, F. M. and Havelaar, A., 'Effect of Accelerated Natural Lactic Fermentation of Infant Food Ingredients on Some Pathogenic Micro-organisms', *Int. Journal of Food Microbiology* 8, 1989a, pp. 351–61

——, Rombouts, F. M. and Hautvast, J. G. A. J., 'Accelerated Natural Lactic Fermentation of Infant Food Formulations', *UNU Food and Nutrition Bulletin* 11 (1) 1989b, pp. 65–73

Nyarko-Fofie, T., 'Small Enterprise Management Development – Subinso no. 1 "Gari" Processing Group – A case study', Paper submitted to Expert Consultation on Small Rural Enterprises in Africa, FAO, Rome, 11–15 December 1989

Obibuaku, L. O., 'Socio-economic Problems in the Adoption Process: Introduction of a hydraulic palm-oil press', *Rural Sociology*, 32, 4, 1967

Rosling, H., 'Cassava Toxicity and Food Security. A review of health effects of cyanide exposure from cassava and of ways to prevent these effects', IITA-UNICEF Programme on Household Food Security and Nutrition, 1987

Schmidt, O. G., 'The Place of Dehulling in African Sorghum and Millet Food Systems', Annual Meeting, GASCA, Munich, 1987

——, 'The Sorghum Dehuller – a case study in innovation', in Marilyn Carr (ed.), *Sustainable Industrial Development*, London, IT Publications, 1988

Svanberg, U. and Lorri, W. S. M., 'Improved Iron Availability in Weaning Foods using Germination and Fermentation', Paper presented at the Third Africa Food and Nutrition Congress, Harare, Zimbabwe, 5–8 September 1988

White, S., 'African Women as Small-scale Entrepreneurs: Their impact on employment creation', in J. Monson and M. Kalb (eds), *Women as Food Producers in Developing Countries*, University of California, 1985

Wiemer, H. J. and Korthals-Altes, F. W., *Small-scale Processing of Oil Fruits and Oil Seeds*, Braunschweig, Vieweg, 1989

Williams, C. E., 'The Effect of Technological Innovation among Rural Women in Nigeria: A case study of "gari" processing in selected villages of Bendel State of Nigeria', in *Women's Contributions to Food Production and Rural Development in Africa*, Proceedings of a Workshop, AAASA, Addis Ababa, 1984

Index

Of Related Interest

The European Experience

Dieter Senghaas

Translated from the German by K. H. Kimmig

A highly original study by a leading scholar of the "North–South Conflict", designed to overcome the presentism in the debate on Third World development and its dilemmas by taking a fresh look at the European experience during the past 150 years. The author's basic question is how far anything can be learned from this experience and from the history of the centre-periphery relationships which emerged in Europe in the period of industrialisation. The book's remarkable scope and historical insights immediately generated a lively debate on its underlying hypotheses and comparative perspectives.

A pathbreaking book on an important subject. This book should fascinate anyone interested in history but it is more important for the present and future of world politics.

Karl W. Deutsch, Professor of Government, Harvard

Published in English in 1985
Available in hardback and paperback editions
hb 0 907582 17 6
pb 0 907582 33 8

Developing Areas

A Book of Readings and Research

Edited by
Vijayan K. Pillai and Lyle W. Shannon

This volume is an interdisciplinary introduction to the social, political, economic and population problems of countries in what is usually referred to as the Third World. More recently these have been defined as the Fourth World, that group of countries which, by reason of their isolation and paucity of resources, have not yet been drawn into the world economy. Although the colonial period is considered as a background to the position of developing areas, emphasis in this volume is placed on the interrelation of major social institutions, the impact of various institutions on economic and social development and, more recently, the impact of uncontrolled development on the ecosystem.

Vijayan K. Pillai is in the Department of Sociology and Social Work, the University of North Texas and **Lyle W. Shannon** is at Iowa Urban Community Research Center, the University of Iowa.

Winter 1991 ca. 450pp. ill., maps, diagr., bibliog., index
0 85496 741 9 cloth ca. £40.00/$72.00